T0314917

Smart Sensors for Environmental and Medical Applications

IEEE Press
445 Hoes Lane
Piscataway, NJ 08854

IEEE Press Editorial Board
Ekram Hossain, *Editor in Chief*

Jón Atli Benediktsson	Bimal Bose	David Alan Grier
Elya B. Joffe	Xiaoou Li	Peter Lian
Andreas Molisch	Saeid Nahavandi	Jeffrey Reed
Diomidis Spinellis	Sarah Spurgeon	Ahmet Murat Tekalp

Smart Sensors for Environmental and Medical Applications

Edited by

Hamida Hallil and Hadi Heidari

IEEE PRESS

WILEY

IEEE Press Series on Sensors
Vladimir Lumelsky, Series Editor

Published by John Wiley & Sons, Inc., Hoboken, New Jersey.
Published simultaneously in Canada.

For general information on our other products and services or for technical support, please contact our Customer Care Department within the United States at (800) 762-2974, outside the United States at (317) 572-3993 or fax (317) 572-4002.

Wiley also publishes its books in a variety of electronic formats. Some content that appears in print may not be available in electronic formats. For more information about Wiley products, visit our web site at www.wiley.com.

Library of Congress Cataloging-in-Publication Data

Names: Hallil, Hamida, 1981– editor. | Heidari, Hadi, editor.
Title: Smart sensors for environmental and medical applications / Hamida
   Hallil, Hadi Heidari.
Description: Hoboken, New Jersey : Wiley-IEEE Press, 2020. | Series: IEEE
   press series on sensors | Includes bibliographical references and index.
Identifiers: LCCN 2020011698 (print) | LCCN 2020011699 (ebook) | ISBN
   9781119587347 (hardback) | ISBN 9781119587354 (adobe pdf) | ISBN
   9781119587378 (epub)
Subjects: LCSH: Biosensors. | Medical instruments and apparatus.
Classification: LCC R857.B54 S64 2020 (print) | LCC R857.B54 (ebook) |
   DDC 610.28/4–dc23
LC record available at https://lccn.loc.gov/2020011698
LC ebook record available at https://lccn.loc.gov/2020011699

Cover Design: Wiley
Cover Image: © toodtuphoto/Shutterstock

Set in 9.5/12.5pt STIXTwoText by SPi Global, Pondicherry, India

Contents

List of Contributors

Mst. Khudishta Aktar
Department of Microbiology and
Hygiene,
Bangladesh Agricultural University,
Mymensingh, Bangladesh

Aadhav Anantharamakrishnan
Centre for Nanotechnology &
Advanced Biomaterials
and
School of Chemical & Biotechnology,
SASTRA Deemed University,
Thanjavur, India

Maria Luisa Braunger
Department of Applied Physics,
"Gleb Wataghin" Institute of Physics,
University of Campinas (UNICAMP),
Campinas, São Paulo, Brazil

Daniele D. Caviglia
COSMIC Lab, University of Genova,
Genova, Italy

Rona Chandrawati
School of Chemical Engineering and
Australian Centre for Nanomedicine
(ACN),
The University of New South Wales
(UNSW Sydney),
Sydney, NSW, Australia

Corinne Dejous
Univ. Bordeaux, CNRS, IMS,
UMR 5218, Bordeaux INP,
F-33405 Talence, France

K. S. Shalini Devi
Centre for Nanotechnology &
Advanced Biomaterials,
SASTRA Deemed University,
Thanjavur, India

Saakshi Dhanekar
Centre for Biomedical Engineering
(CBME),
Indian Institute of Technology (IIT),
New Delhi, India
and
Department of Electrical Engineering,
Indian Institute of Technology Jodhpur,
Karwar, Rajasthan, India

Hamida Hallil
Univ. Bordeaux, CNRS, IMS,
UMR 5218, Bordeaux INP,
F-33405 Talence, France
and
CINTRA, CNRS/NTU/THALES,
UMI 3288, Research Techno Plaza,
Singapore 637553, Singapore

Hadi Heidari
School of Engineering,
University of Glasgow,
Glasgow, UK

Zeinab Hijazi
COSMIC Lab, University of Genova,
Genova, Italy
and
Department of Electric and Electronic
Engineering,
International University of Beirut (BIU),
Beirut, Lebanon

Md. Abdul Kafi
Department of Microbiology and
Hygiene,
Bangladesh Agricultural University,
Mymensingh, Bangladesh

Uma Maheswari Krishnan
Centre for Nanotechnology &
Advanced Biomaterials,
School of Chemical & Biotechnology,
and
School of Arts, Science & Humanities,
SASTRA Deemed University,
Thanjavur, India

France Le Bihan
CNRS, IETR, University of Rennes 1,
Rennes, France

Federico Mazur
School of Chemical Engineering and
Australian Centre for Nanomedicine
(ACN),
The University of New South Wales
(UNSW Sydney),
Sydney, NSW, Australia

Muis Muthadi
Institute of Innovative Research,
Tokyo Institute of Technology, Japan

Takamichi Nakamoto
Institute of Innovative Research,
Tokyo Institute of Technology, Japan

Osvaldo N. Oliveira, Jr.
São Carlos Institute of Physics,
University of São Paulo (USP),
São Carlos, São Paulo, Brazil

Laurent Pichon
CNRS, IETR, University of Rennes 1,
Rennes, France

Antonio Riul, Jr.
Department of Applied Physics,
"Gleb Wataghin" Institute of Physics,
University of Campinas (UNICAMP),
Campinas, São Paulo, Brazil

Anne-Claire Salaün
CNRS, IETR, University of Rennes 1,
Rennes, France

Flavio M. Shimizu
Brazilian Nanotechnology National
Laboratory (LNNano),
Brazilian Center for Research in
Energy and Materials (CNPEM),
Campinas, São Paulo, Brazil

Maurizio Valle
COSMIC Lab, University of Genova,
Genova, Italy

Max Weston
School of Chemical Engineering and
Australian Centre for Nanomedicine
(ACN),
The University of New South Wales
(UNSW Sydney),
Sydney, NSW, Australia

Jatinder Yakhmi
Bhabha Atomic Research Centre,
Mumbai, India

Preface

The future of research in the field of bio and chemical sensors towards obtaining smart systems is challenging and enchanting. It nonstop restores itself in respect to market demands, technological innovations and advancements in novel sensing materials. Key interests driving the bio-chemical sensors market across the world are occupational safety health, hazards detection, environment pollution control, air quality analysis, food safety and quality, healthcare device and biomedical applications. Because of rapid evolution of smart bio-chemical sensors, its market in exploiting such technology in wearable devices and internet of things predicts a gigantic growth over next few years. Large and ever-growing advances in developing and implementing such technologies exhibited the potential utility of this unique class platforms as future of environmental and medical sensing systems. This book through various chapters will help readers gain insight into the technical problems that must be overcome while developing complex and smart bio-chemical sensors. The rationale for selecting these paradigms is provided by a number of breakthroughs, which have been recently achieved in this field.

This book presents a comprehensive overview of bio-chemical sensors, ranging from the choice of material to sensor validation, modeling, simulation, and manufacturing. It discusses the process of data collection by intelligent techniques such as deep learning, multivariate analysis, and others. It also incorporates different types of smart chemical sensors and discusses each under a common set of subsections so that readers can fully understand the advantages and disadvantages of the relevant transducers—depending on the design, transduction mode, and final applications.

The book targets post-graduate students and young researchers as well as engineers working in industry willing to understand and connect bio-chemical sensors with state-of-the-art and emerging medical and environmental applications. The field of biomedical electronics spanning from biology and device technology to smart sensors are covered with emphasis on smart bio-chemical systems. The contents ensure a good balance between academia and industry, combined with a judicious selection of distinguished world-leading authors.

This book addresses the limitations and challenges in the state-of-the-art smart bio-chemical sensors. It includes ten chapters of contributions from leading experts in bio and chemical sensing. We believe that the approaches developed, and the issues raised in this book will enable the reader to identify the requirements, challenges and future directions related to the burgeoning field of bio-chemical detection systems.

In **Chapter One** we will to recall and explain some basic principles and metrological characteristics common to various sensors categories. These basic notions will provide the reader with a foundation and knowledge for understanding the different technologies and issues raised in the presented chapters. Different state-of-the-art Field Effect transistors (FETs) sensors and their applications, especially for bio-chemical detection have been described in **Chapter Two**. It presents various stages in the development of the sensors based on FETS which benefit from advanced research trends, especially the development of new materials, compatible with low-cost technologies on various substrates, including flexible substrates. **Chapter Three** evaluates performances of cell based electrochemical methods for screening cells of different origin and cell from the specific stages of a cell cycle. In addition, this chapter discusses about the conformability and biocompatibility of the cell-based platform essential for wearable and implantable application. **Chapter Four** introduces Electronic tongues (e-tongues) that are promising electroanalytical devices for the quality control of water, beverages, foodstuffs, pharmaceuticals and complex liquids as they offer simple operation, fast response, low cost, and high sensitivity. **Chapter Five** discusses recent developments of colorimetric sensors based on polydiacetylene and liposomes for the detection of food spoilage, specifically their detection mechanism, sensitivity and specificity towards analytes in food. In **Chapter Six** concerted efforts are underway to bring down operating temperature of the MOx sensors. Tailoring the size and shape of nanostructured MOxs to tune their gas sensing properties is a major area of research and has opened up new vistas in analytical chemistry and instrument engineering. **Chapter seven** will first provide the main requirements for a gas sensing system and a general introduction about the types of chemosensors highlighting the advantages and drawbacks: the focus will be on Metal Oxide (MOX) gas sensors for their high sensitivity, fast response time, long lifetime and small size. It will then introduce the state-of-the-art of sensor circuit interfaces towards the implementation of compact, reliable, low cost, low power e-nose systems. **Chapter Eight** gives an insight of the E-nose, structure with its components like the sensor array and the pattern recognition methods. In **Chapter Nine** recent hot topics such as odor biosensor, prediction of odor impression and strategy for odor-source localization are described. The **last Chapter** presents a state of the art enabling to position the transducers based on microwave transduction. It will follow by a review on the different designs of developed platforms and the associated sensitive materials.

We hope that this book will serve as a useful resource to researchers and scientists in academia as well as industry in their effort to turn the new paradigm of emerging smart bio-chemical sensing technology and application into biomedical sensing systems and to overcome bottlenecks in bio-chemical sensors development.

We are grateful to all authors who have contributed their time and energy to make this book a reality. In their name I also thank those people within their organizations who provided assistance to them. The compilation and editing of this book were, with great enthusiasm, supported by IEEE Sensors Council and Wiley-IEEE Press.

Hamida Hallil, Hadi Heidari

About the Editors

Hamida Hallil, PhD, is an Associate Professor in Electrical Engineering at the Bordeaux University and affiliated with the laboratory of Integration: From Material to Systems. Her current research interests include the design of innovative devices and sensors using electromagnetic and acoustic transduction modes. Since 2018, she is assigned as research scientist at CNRS International-NTU-Thales Research Alliance in Singapore and her work focuses on the development of 2D-based acoustic devices and microwave sensors. She has coauthored over 60 peer-reviewed journal articles and conferences. She serves on the organizing or technical committees of several international conferences and French organizations.

Hadi Heidari is an Assistant Professor (Lecturer) in Electronics and Nanoscale Engineering and lead of meLAB at the University of Glasgow, UK. His research focuses on microelectronics and sensors for wearable and implantable devices. He has authored over 140 articles in top-tier peer-reviewed journals and in international conferences. He is an IEEE Senior Member, an Associate Editor for four journals, and the General Chair of IEEE ICECS 2020 Conference. He is member of the IEEE Circuits and Systems Society Board of Governors and Member-at-Large in IEEE Sensors Council. He has grant portfolio of +£1 million funded by major research councils and funding organizations including the European Commission, UK's EPSRC, Royal Society, and Scottish Funding Council.

1

Introduction

Hamida Hallil[1,2] and Hadi Heidari[3]

[1] *Univ. Bordeaux, CNRS, IMS, UMR 5218, Bordeaux INP, F-33405 Talence, France*
[2] *CINTRA, CNRS/NTU/THALES, UMI 3288, Research Techno Plaza, Singapore 637553, Singapore*
[3] *School of Engineering, University of Glasgow, Glasgow, UK*

1.1 Overview

Scientific and technological advances of recent years allow considering the real-time detection of toxic pollutants or chemical or biological substances in gaseous or liquid environments adequately. It is possible to easily find on the market portable devices that allow, for investments of a few hundred to a few thousand dollars, sensor or diagnostic platforms, or low concentrations of chemical or biological species. A smooth, fast, and cost-effective detection of the presence of a chemical or biological element and the quantification of its concentration in real time are criteria that help to amplify the distribution of these sensors and access to highly sought-after measurements particularly in demanding areas of scientific knowledge at the boundaries between applied mathematics, physics, chemistry, and biology. This enthusiasm is particularly noticeable in applications dedicated to the issues from the environment, food, and health.

Nowadays, the various commercialized systems existing to answer these issues can be presented in two different approaches: the sensors dedicated to the identification of the risks and consequently to alarm the user; and sensors dedicated to the specific detection of target species at very low concentrations in real time.

However, despite these remarkable technological advances, the development of sensors with: (i) high sensitivity, (ii) real selectivity to a biological or chemical species, (iii) low limit of quantification, (iv) energy autonomy, and (v) reasonable cost remain ultimate challenges for manufacturers and academic researchers.

Smart Sensors for Environmental and Medical Applications, First Edition. Edited by Hamida Hallil and Hadi Heidari.
© 2020 The Institute of Electrical and Electronics Engineers, Inc.
Published 2020 by John Wiley & Sons, Inc.

In recent years based on academic literature, an enormous surge of works has been carried out to develop robust, reliable, accurate, and high-resolution chemical sensing platforms. Also, many efforts have been attempted to convert them into miniaturized, more portable, and cost-effective systems and to study protocols currently used in advanced sensor networks.

A surge of interest, yet an unmet market demand for reliable and high-performance chemical sensors from different perspectives from materials (polymer, metal oxide, carbon material, etc.) and technology (electrochemical, Field Effect Transistor [FET], acoustic, microwave, optic, electronic tongue and nose, etc.) to applications (food spoilage monitoring, odor, medical, environmental, IOT, etc.), and the accurate interpretation of biochemical processes by readily measurable signals still exists. Such biochemical sensors need to provide fast response, high-sensitivity and selectivity, large dynamic range, and low-cost to be considered as viable products. These sensors can serve as various applications such as biothreat detection, epidemic disease control, low-cost home healthcare, and cell-based and environmental monitoring.

This book titled *Smart Sensors for Environmental and Medical Applications* addresses the limitations and challenges in obtaining the state-of-the-art smart biochemical sensors. It includes ten chapters of contributions from leading experts in bio and chemical sensing. We believe that the approaches developed, and the issues raised in this book will enable the reader to identify the requirements, challenges, and future directions related to the burgeoning field of biochemical detection systems. It should be noticed that in this introduction it is important to recall and explain some basic principles and metrological characteristics common to various sensors categories. These basic notions will provide the reader with a foundation and knowledge for understanding the different technologies and issues raised in the presented chapters.

Furthermore, this book will allow the readers to identify new opportunities in this emerging research field.

1.2 Sensors: History and Terminology

Scientific knowledge has developed through a double effort:

- First, the reflection on the mechanisms, that is to say on the nature of interactions between physical and chemical quantities-related phenomena; this thinking is reflected by the mathematical tool by the laws of physics, abstract relationships between physical quantities.
- Second, experimentation based on the measurement of physical and chemical quantities and which, by associating a numerical value allows to quantitatively define the properties of objects, digitally verify the physical laws, or to empirically establish the form.

Whereas science seeks to grasp and then to express coherent mathematical theories and the laws governing the relationships of physical quantities, technology uses these laws and the properties of matter to develop new devices or materials that enable humans to increase their means of action to better support their wellbeing, facilitate their exchanges, and improve their life. Indeed, at first, the technique was a collection of experimental processes, fruits of the observation, random groupings, or successive tests; the knowledge of the laws of nature allowed the technique to rationalize its approach and to become a science of realization. The measure therefore plays a crucial role. In order to be carried out successfully, the measuring operation generally requires that the information be transmitted remotely from the point where it is captured, protected against alteration by parasitic phenomena, and amplified, before being operated in various ways: displayed, saved, and processed by calculator.

In this respect, electronics offer a variety of influential means: to benefit from measurements of all types of physical quantities, such as their processing and exploitation, it is very desirable to transpose each of the physical quantities immediately into the form of an electrical signal. It is the role of the sensor to ensure this duplication of information by transferring it, at the very point where the measurement is made, of the physical quantity (nonelectric) of its own, on an electrical quantity: current, voltage, load, or impedance.

A sensor is first of all the result of the ingenious exploitation of physical law: this is why an important place is given in this book to the physical principles which are at their base. This is the result of specific properties of each type of sensor: performance, field of application, and rules of good use.

The electrical characteristics of the sensor impose on the user the choice of associated electrical circuits that are perfectly adapted. Therefore, the delivered signal is obtained and can be processed under the best conditions. Indeed, physical principles, specific properties, and associated electrical assemblies are the three main aspects under which each type of sensor will be studied.

1.2.1 Definitions and General Characteristics

The physical quantity that is the object of measurement (temperature, pressure, magnetic, humidity, gas molecules, biomarker, deformation, etc.) is designated as the measurand and represented by M; all the experimental operations which contribute to the knowledge of the numerical value of the measurand constitute its measurement [1].

The sensor is a device that is subjected to the action of a physical or chemical phenomenon measurand, which has a characteristic of electrical nature (load, voltage, current, or impedance) designated by R and which is a function of the measurand: $R = F(M)$ (Figure 1.1). R is the response or the output quantity of the sensor. The measurement of R should allow to know the value of M.

The relation $R = F(M)$ results in its theoretical form from the physical laws which govern the operating of the sensor and in the numerical expression of its design (geometry, dimensions), of the materials which constitute it and possibly of its environment, and its mode of use (temperature, power supply).

For any type of sensor, the relation $R = F(M)$ in its numerically exploitable form is explained by calibration: for a set of precisely known values of M, we measure the corresponding values of R which makes it possible to draw the curve of calibration (Figure 1.2a). The calibration curve, at any measured value of R, makes it possible to associate the value of M which determines it (Figure 1.2b).

For reasons of operation ease, efforts are made to make the sensor, or at least to use it, so that it establishes a linear relationship between the variations ΔR of the output quantity and those ΔM of the input quantity: $\Delta R = S\Delta M$. S is the sensitivity of the sensor.

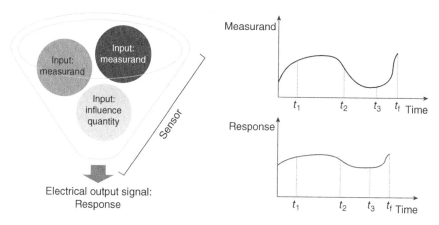

Figure 1.1 Example of evolution of a measurand M and the corresponding response R of the sensor. *Source:* Adapted from [1].

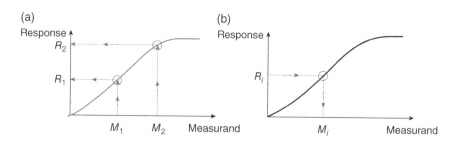

Figure 1.2 Calibration curve of a sensor: (a) its establishment, based on known values of the measurand M; (b) its exploitation, from the measured values of the sensor response R. *Source:* Adapted from [1].

One of the major challenges, in the design and use of a sensor, is the constancy of its sensitivity S which must depend as little as possible on:

- the value of M (linearity) and its frequency of variation (bandwidth);
- time (aging);
- disturbances of other physical quantities of its environment which are not the object of measurement and which are designated as quantities of influence.

As an element of electrical circuit, the sensor is presented, seen from its output:

- either as a generator, R being a load, a voltage, or a current and it is then an active sensor;
- or as an impedance, R then being a resistance, an inductance, or a capacitance: the sensor is then said to be passive.

This distinction between active and passive sensors based on their equivalent electrical circuits reflects a fundamental difference involved in nature of the physical phenomena.

1.2.2 Influence Quantities

The sensor, by its conditions of use, may be subject not only to the measurand but also to other physical quantities whose variations have potential to cause a change in the output electrical magnitude that it is not possible to distinguish from the action of the measurand. These "parasitic" physical quantities to which the response of the sensor can be sensitive are the influence quantities. Thus, we can mention:

- the temperature, which modifies the electrical, mechanical, and dimensional characteristics of the sensor elements;
- the pressure, the acceleration, and vibrations have potential to create in certain sensors parts deformations and constraints that alter the response;
- the humidity at which certain electrical properties such as dielectric constant or resistivity can be sensitive and which may degrade the electrical insulation between sensor components or between the sensor and its environment;
- variable or static magnetic fields: the first create electromotive force and the second can change an electrical property, such as resistivity, when the sensor uses a magnetoresistive material;
- the supply voltage.

In order to be able to deduce the value of M from the measurement of R, it is therefore necessary:

- to reduce the importance of influence quantities on the sensor by protecting it with adequate insulation: antivibration mounts, magnetic shielding;

- either to stabilize the influence quantities to perfectly known values and to calibrate the sensor under these operating conditions: thermostatically controlled, regulated power sources;
- or finally to use circuits that allow to compensate the influence quantities: Wheatstone bridge, differential measurements, etc.

All these notions will be taken up and treated in depth in examples illustrated in the chapters of this book.

1.3 Smart Sensors for Environmental and Medical Applications

Smart sensors in environmental and medical applications are intended for the detection and/or analysis of:

- changes in physical parameters (vibration, temperature, and pressure) and
- the concentration of chemical/biochemical or biological species, gaseous or liquid in general. The field relating to biochemistry and biology is extremely broad and therefore involves a large number of application areas: the chemical industry, fine chemicals, cosmetics, pharmaceuticals, health, agri-food, the environment, home automation, civil security, etc.

In general, it is possible to define a smart sensor in environmental and medical applications by the following elements:

- a medium to be analyzed (temperature, pressure, gases, liquids, tissues, etc.) with its specificities;
- a detection/transduction principle for transforming the analyzed physical/chemical signal into a measurable electrical quantity;
- a measurement and signal processing interface for shaping the useful electrical signal;
- data processing coupled with a calibration system to ensure the reliability of the measurement;
- a source of energy to ensure the autonomy of the whole chain of detection/transduction/data processing.

Different from other types of physical sensors, chemical/biological sensors in environmental and medical applications have had limited commercial success, and this despite the strong demand of many sectors of the economy cited above. This fact is probably linked to the technical difficulty of transforming a quantity such as the concentration of species in a liquid or a gas into an electrical signal, while simultaneously ensuring reproducibility, sensitivity, and selectivity.

The various types of chemical/biological sensors existing and treated in the chapters presented in this book operate according to very varied physicochemical principles in environmental and medical applications.

The term smart or intelligent sensor refers to an instrument in digital technology combining data acquisition and their internal processing and incorporates new features. In general, these sensors integrate an embedded microcontroller to perform internal processing and calculations and have a bidirectional communication capability that means receiving external commands and sending measurements and status information. They are equipped with one or more sensor matrices for measuring the target and influence quantities and other integrated algorithms for the analysis of these measurements and therefore provide decision support.

With the current technological advances in data collection, storage, and processing, it is therefore natural that robust solutions such as combinations of multivariate data analysis and automatic learning methods are associated to biological and chemical sensor arrays to improve their performance in terms of selectivity and identification of target species in complex environments. These methods are widely and successfully utilized for classification in several fields and exhibit high performance.

Human and animals are capable of sensing different types of smell in the vicinity by a natural nose. Scientists investigate "an artificial nose," which had to be trained and then could smartly sense smell like humans; it was called "Electronic Nose" or in short "E-Nose." It has wide applications in the field of smart biological and chemical sensors, of which toxic gas detection is one of the key areas. Another critical application where such sensors will be useful is "healthcare." An E-Nose is an association of advanced algorithms based in multivariate data analysis or machine learning and a variety of transducers dedicated for detection and identification of odorants.

Over the last decade, another kind of biological and chemical sensors has been emerged: Electronic Tongues (e-tongues). E-tongues are promising electroanalytical devices for the quality control of water, beverages, foodstuffs, pharmaceuticals, and complex liquids as they offer simple operation, fast response, low cost, and high sensitivity and selectivity. They comprise an array of sensing units having distinct responses to establish a fingerprint of the samples, being based on the global selectivity concept. In mimicking biological systems, ETs serve to classify large amounts of information into specific patterns. The integration of ETs with microfluidic chips expands potential applications due to miniaturization, and usage of microliters for sampling and discharge, crucial when hazardous reagents or biological materials are studied. The concept of e-tongue was also extended to biosensing, with biomolecules used as sensing units reaching the recognition ability at the molecular level.

The chapters presented in this book provide an in-depth discussion of the necessary definitions of the different transducers associated with these technologies, their metrological performance, associated electronic systems for data processing, and classification methods and associated analyses.

1.4 Outline

We know that only measurements and experiments can drive scientific progress and increase knowledge of what surrounds us; the limits are not reached and never will be. It seems to us, to this day, that the measurement plays a fundamental role in the development or the follow-up of the industrial and human activities and the technologies of the future with the sophistication of automatisms, robotics, the control of the quality, energy, pollution control, disease diagnosis or drug screening, etc. In addition, the measurement, through emerging areas such as artificial intelligence and the Internet of Things and biotechnologies or even neuroscience, is now finding many applications in the design and realization of the systems of the future.

Measurement becomes an essential factor of the economy; it must be treated with special and sustained attention and nothing will be done without the "sensor" at the cornerstone "measure."

This book incorporates different types of smart sensors and discusses each of the chapters in light of a common set of sub-sections so that the readers can be educated about the advantages and disadvantages of the relevant transducers depending on the design, transduction mode, and applications. The book covers all of the major aspects of the primary constituents of the field of smart sensors for environmental and medical applications including working principle and related theory, sensor materials, classification of respective transducer type, relevant fabrication processes, methods for data analysis, and suitable application. In the theory, the book discusses the fundamentals of the sensing phenomena and the relevant equation governing the phenomena based on selected transduction mode. The section on sensor materials not only discusses the promising materials whose properties have been utilized for sensing action but also predicts future innovative materials that have the potential for sensing application. The classification section categorizes the sensor into different sub-types and describes their working, focusing on prominent applications for the readers to realize the benefits of relevant designs and selected methods for data analysis. The application section includes the state-of-the-art update of the developments in the field of the given sensor type and concludes with the challenges in the relevant field of research. Different transduction modes which have been applied in the design and fabrication of various biological and chemical sensors are discussed in the book, as illustrated in Figure 1.3. The massive amounts of data generated in experiments with

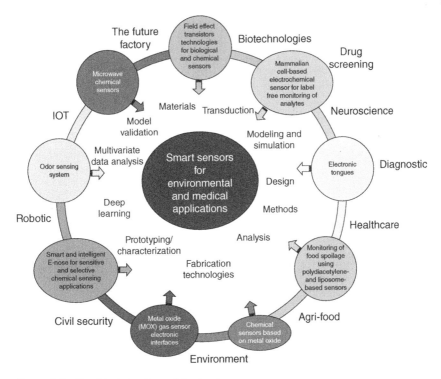

Figure 1.3 Graphical structure of the book.

sensors have motivated studies involving multidimensional projection techniques, feature selection, and machine learning.

The most recently developed state-of-the-art sensors, and still being tested in leading research laboratories, are provided and discussed. The measurement specialists or students will be surprised to discover, on reading, smart chemical sensor novelties that have not yet come to this knowledge. The recent technologies in the smart chemical sensing would fulfil the demand for a book, which has a balance between fundamentals of sensor design, fabrication, characterization, and analysis combined with emerging methods (e.g. machine learning). Furthermore, it should benefit a wide range of students from graduates to undergraduate as well as incorporates recent developments in research that would suit doctoral students, postdoctoral fellows, and industrial engineers.

Reference

1 Asch, G. and Poussery, B. (2017). *Les capteurs en instrumentation industrielle*, 8e. Dunod.

2

Field Effect Transistor Technologies for Biological and Chemical Sensors

Anne-Claire Salaün, France Le Bihan, and Laurent Pichon

CNRS, IETR, University of Rennes 1, Rennes, France

2.1 Introduction

Field Effect transistors (FETs) are powerful transducers for chemical and biological detection and offer a great promising method for label-free, ultrasensitive, and real-time selective detection of charged species. They are compatible with industrial fabrications and allow low manufacturing cost and power consumption in addition to the advantageous electronic features of embedded detection and signal processing in silicon technology. This transduction method resulting in conductance variation due to the charges localized on the sensitive layer promotes high sensitivity with low detection time. The sensitive layers can be easily functionalized in order to address many different applications.

These sensors based on FETS benefit from advanced research trends, especially the development of new materials, compatible with low-cost technologies on various substrates, including flexible substrates. Device advancements increase their performances and reliability. They are compatible with gas or liquid media, thus enhancing their possible applications, either as gas detection, chemical sensing, bioelement recognition, dosing, or microfluidic integration.

In this chapter, different state-of-the-art FET sensors and their applications, especially for biochemical detection, are reviewed.

Smart Sensors for Environmental and Medical Applications, First Edition. Edited by
Hamida Hallil and Hadi Heidari.
© 2020 The Institute of Electrical and Electronics Engineers, Inc.
Published 2020 by John Wiley & Sons, Inc.

2.2 FET Gas Sensors

2.2.1 Materials

2.2.1.1 Inorganic Semiconductors

The most widespread sensors are those based on semiconductors (mainly metal oxides), because even if some problems in use exist from the point of view of selectivity and stability, they have many advantages such as their sensitivity, response time, price, and integration (Table 2.1). Among the most used, one can mention tin oxide (SnO_2), zinc oxide (ZnO_2), copper oxide (CuO), and tungsten trioxide (WO_3). They target the measurement of gases such as carbon monoxide or nitrogen dioxide responsible for an oxidation-reduction reaction with the oxygen atoms of the metal oxide. This reaction causes a decrease in the electrical resistance of the semiconductor layer. Many variants of gas sensors based on metal oxides have been developed in the last decade [1–8].

2.2.1.2 Semiconductor Polymers

Since the mid-1990s and mainly the early 2000s, many researchers have focused on the use of conductive polymers as a sensitive material. Globally, there are four main ones: Polyaniline (PAni), Polypyrrole, Poly 3,4-ethylenedioxythiophene, and Phthalocyanines.

Polymers are in great demand in the field of sensors because of their large number and diversity [9–12] and also because their physicochemical properties can be easily adapted by adding additives or new monomers to obtain new copolymers. Another advantage is that the reactions with the gas molecules take place at room temperature [13]. Polymers also stand out thanks to their flexibility and low energy consumption. However, their properties are not stable over time and therefore the performances of the sensors are degraded.

Table 2.1 Gas sensitivity of main metal oxide.

Gas material	CO	NO$_x$	CH$_4$	NH$_3$	H$_2$	SO$_2$	C$_2$H$_4$O	H$_2$S	C$_3$H$_8$	Alcohols
CuO	*		*				*			
Fe$_2$O$_3$	*	*	*							
In$_2$O$_3$	*	*	*	*						
SnO$_2$	*	*	*	*	*	*	*		*	
TiO$_2$	*			*		*		*		
WO$_3$		*				*		*	*	*
ZnO	*	*	*	*	*		*			*

2.2.1.3 Nanostructured Materials

For gas detection, nanomaterials are excellent transducers and can improve performance such as sensitivity, response time, selectivity, stability, energy consumption, and ambient temperature monitoring. Because of their small size with a high surface area to volume ratio, some gas molecules can modify the properties of the material, and the detection of very low concentration in a short period of time is then possible. For example, the nanowire-based gas sensors seem to be a more favorable configuration than the traditional macroscopic sensors, for the co-integration of several different sensors on the same chip. So, the selectivity and stability of the gas detection devices can be improved. In addition, since very small amounts of gas could change the characteristics of the nanowires at room temperature, this would allow sensors to work at low power. The use of nanostructure would lead to reduced size of the sensors, causing a loss of the weight of the component, minimizing energy consumption, with objective for integration in embedded systems.

2.2.2 FET as Gas Sensors

Despite a more complicated process manufacturing of resistors used as sensors, the use of FETs as sensors has been more widespread, thanks to higher sensitivity.

2.2.2.1 Pioneering FET Gas Sensors

Pioneering gas FETs were issued from classical complementary metal oxide semiconductor (CMOS) technology with the gate electrode covered by a sensitive material such as palladium for hydrogen detection with a sensitivity of 10 ppm and reasonable response times at 150 °C [14]. Since, Suspended Gate FET (SGFET) gas sensors have been developed depending on design and sensitive material. The gas can reach the sensitive surface through the air gap without having to diffuse through the gate electrode. The devices also do not require a high temperature to work. A wide range of sensitive materials (metal, oxide, inorganic, or organic) was investigated for selective detection of nitrogen dioxide, hydrogen, ammonia, and carbon monoxide at room temperature [15]. An alternative structure of SGFET was also proposed as capacitively controlled field effect transistor (CCFET) [16]. In this structure, a capacitor with an air gap covered by a sensitive layer controls the gate electrode potential of the readout FET (Figure 2.1).

Recent advances in semiconductor synthesis methods (organic and inorganic) have led to the emergence of gas sensors based on new polymeric materials, and also on nanomaterials.

2.2.2.2 OFET Gas Sensors

The variety of organic transistors used in the detection area is constantly increasing because of their properties. The structure and morphology of the molecules can be

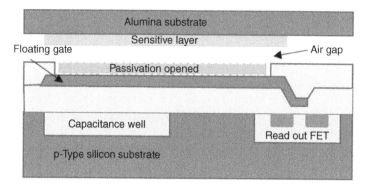

Figure 2.1 Schematic illustration of a CCFET gas sensor.

adjusted to improve the sensitivity and selectivity of the gas sensor. In addition, the rapid development of organic electronics and the improvement of organic field effect transistors (OFETs) make them highly suitable for large-scale manufacturing, and attractive for miniaturized gas detection system.

Semiconducting polymers most commonly used for OFET gas sensors are polymethyl methacrylate (PMMA) as gate insulator, and Pentacene (e.g. ammonia detection) or phthalocyanine (detection of NO_2) as sensitive active layers (Figure 2.2) [17–19].

In order to improve the performance of OFET-based sensors, efforts have been made to design the structures [17, 20], or using a heterojunction-sensitive layer [21], or by incorporating nanoparticles into the structure (Figure 2.3) [18]. In particular, the use of incorporated ZnO nanoparticles into a PMMA gate dielectric can facilitate NH_3 detection with a higher response time.

2.2.2.3 Nanowires-Based FET Gas Sensors

Transistor based on nanowires can be an alternative structure to the classical FET. Silicon [22–24] or metal oxides [25] are the most materials used as nanowires. Compared to the classical FET configuration, the nanowire plays the role of channel, addressed at the ends by the source and the drain electrodes. In most cases the assembly is made on a thin layer of oxide, on a substrate conductor used as gate electrode (Figure 2.4). When the surface of the nanowire is exposed to the gas, a specific reaction leads to a change in conductance of the nanowire.

The FET configuration is more sensitive than the resistors due to its ability to modulate nanowire conductance through the gate. However, the manufacturing process is more complex in particular because of the manufacture and control of the gate on the rear face.

Liao et al. propose sensors using CuO nanowires as a conduction channel in FETs [26]. These sensors have demonstrated a high and fast response time compared to planar devices, under exposure of CO, and ethanol in less counterpart.

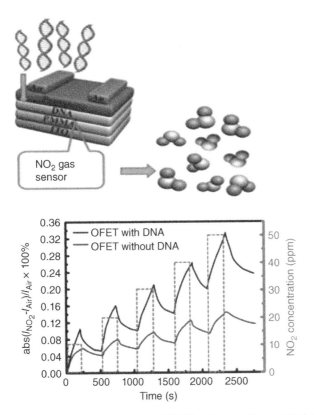

Figure 2.2 Response curves of OFET with and without DNA interlayer after stored in atmosphere for 30 days upon exposure to NO_2 in different concentrations. *Source:* From Shi et al. [17], reprinted with permission.

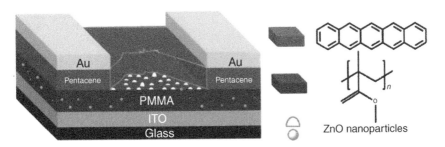

Figure 2.3 Schematic structure of OFET sensor based on ZnO/PMMA hybrid dielectric and molecular structures of pentacene and PMMA. *Source:* From Han et al. [18], reprinted with permission.

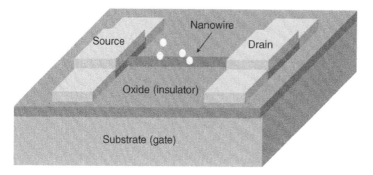

Figure 2.4 FET transistor based on nanowire.

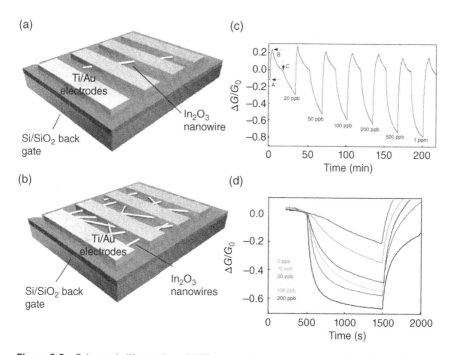

Figure 2.5 Schematic illustration of FET sensor based on one nanowire (a), multiple nanowires (b). Sensors' electrical responses under NO_2 exposure with one nanowire (c), multiple nanowires (d). *Source:* From Zhang et al. [27], reprinted with permission.

Zhang et al. developed nanowire-based transistors of In_2O_3 for detecting NO_2 concentrations at ppb level at room temperature [27]. The single nanowire sensors of In_2O_3 showed a limit of detection of approximately 20 ppb, whereas the configuration with multiple nanowires showed an even lower limit of detection ($\approx$5 ppb) (Figure 2.5). However, the sensitivity of the sensor remains independent of the number of nanowires (0.5–20 ppb).

Another work [28] proposes a method for transferring hundreds of prealigned nanowires onto a flexible plastic substrate, to produce low power, gas-sensitive, ordered nanostructure films (Figure 2.6a). In this case the nanowires FET demonstrates sensitivity for very low NO_2 concentrations (of the order of the ppb) (Figure 2.6b). However, the response time obtained is long, suggesting that the limit of this device is reached. On the other hand, the reversibility of the sensor is

(a)   (b)

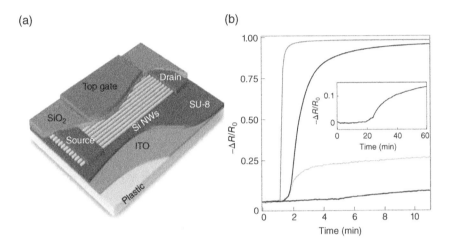

(c)

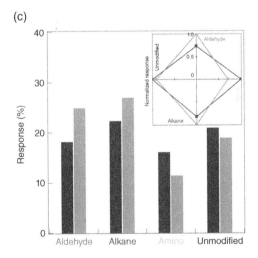

Figure 2.6 (a) Schematic illustration of FET sensor based on multiple nanowires on plastic substrate. (b) Electrical response of the sensor under NO_2 exposure at different concentrations, 20 ppm, 2 ppm, 200 ppb, and 20 ppb. (c) Bar graph of the response to acetone (black) and hexane (grey) exposure. *Source:* From McAlpine et al. [28], reprinted with permission.

Table 2.2 Nanowires-based GasFET sensing properties for ZnO, InO$_3$, and Si materials.

Material	Target gas	Media	Detection limit	Response time	Reference
ZnO	NO$_2$	Ar	1 ppm	N/A	[30]
	NH$_3$		0.5%		
In$_2$O$_3$	NH$_3$	Air	0.02%	2 min	[31]
	NO$_2$		5 ppb	15 min	[27]
Si	NO$_2$	N$_2$	20 ppb	>3 min	[28]

not natural and requires particular conditions to be obtained (UV illumination). In addition, the same group has demonstrated that silicon nanowires can be easily functionalized to discriminate different gaseous environments (Figure 2.6c) [28, 29]. Table 2.2 shows summed up nanowires-based Gas FET with sensitive ZnO, InO$_3$, and Si materials, detecting NO$_2$ and NH$_3$ gas.

2.3 Ion-Sensitive Field Effect Transistors Based Devices

2.3.1 Classical ISFET

As previously shown, FETs can be used as interesting transducers for the detection of charges linked to their surface, with different possible configurations. For ionic detection, but also especially in the case of biological elements, measurement must be achieved in liquid media. Bergveld [32, 33] first adapted the FET configuration for suitable measurement in liquids. The structure of the Ion-Sensitive Field Effect Transistor (ISFET) is based on Metal Oxide Semiconductor Field Effect Transistor (MOSFET) structure, but the gate contact is replaced by the solution and a reference electrode as shown in Figure 2.7a. Charges linked to the surface produce a shift of the threshold voltage of the transistor (Figure 2.7b). This shift is proportional to the log scale of charges concentration. Many theoretical studies of the chemical reaction with surface, the double layer formation in the solution, and the modeling of the sensitivity of such devices were proposed [34, 35]. The main application, currently available and commercialized, concerns the measurement of the pH of solutions, for which the sensitivity of the device is constant in all the usual pH ranges. This sensitivity, as theoretically demonstrated [34–36], is limited by Nernst equation, to 59 mV/pH (see Figure 2.7c).

Many studies were performed in order to optimize the linearity of the response as well as the sensitivity, especially by testing different gate insulator materials, and also for the integration of a pseudo reference electrode required for the liquid polarization.

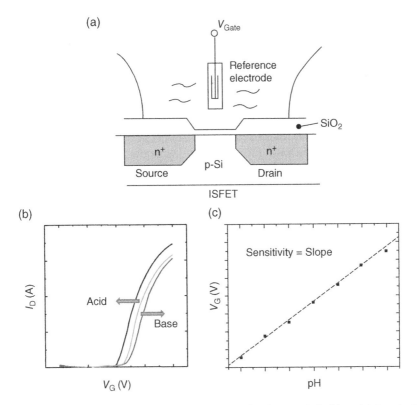

Figure 2.7 ISFET structure (a), shift of the transfer characteristic (b), and determination of the pH sensitivity (c).

Many applications were developed not only for chemical detection with specific layers (ChemFET), but also for biological detections with specific functionalizations.

The compatibility of these devices to electronic circuits [37] and microfluidic integration allowed the development of many applications not only for environments as water analyses [38], soil [39, 40], or pesticides [41], but also for health, food industry, safety, and so on.

2.3.2 Other Technologies

Different technologies are based on ISFET structure, with different goals, as for instance a simpler use with better performances (ageing, shift...), or new applications with new materials (organic materials) or in order to increase the sensitivity of the detection.

2.3.2.1 EGFET: Extended Gate FET

This technology is different from the ISFET one's and have a real advantage of separating the FET device from the chemical environment [42–44]. It is composed of a MOSFET that can be a commercial one, linked to a sensitive membrane by a conductive material, in order to create an extended gate contact.

This technology has many advantages as it is easier to passivate and package, and the extended gate area, with no constraint on its size and shape, is highly compatible with microfluidic integration. The transducers, usually MOSFET, can be fabricated by various industrial technologies or integrated on the same substrate. One example is given in Figure 2.8. The sensitivity is in this case comparable to that of ISFET.

2.3.2.2 SGFET: Suspended Gate FFETs

This technology, previously described for gas sensing, is comparable to ISFET structure (Figure 2.9), associated to a suspended gate insulated from the liquid media. The chemical or biological molecules can be linked on the sensitive surface below the suspended gate. This technology doesn't require a reference electrode and was used for pH, DNA, or protein detection.

2.3.2.3 DGFET: Dual-Gate FETs

A new technology involving a dual gate was recently developed for chemical and biological sensing. Thanks to a capacitive coupling, the sensitivity of such devices

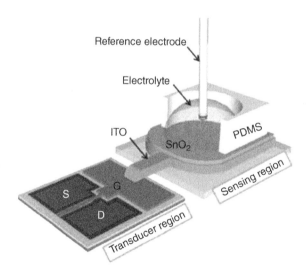

Figure 2.8 Example of extended gate FET structures. *Source:* From Park and Cho [45], reprinted with permission.

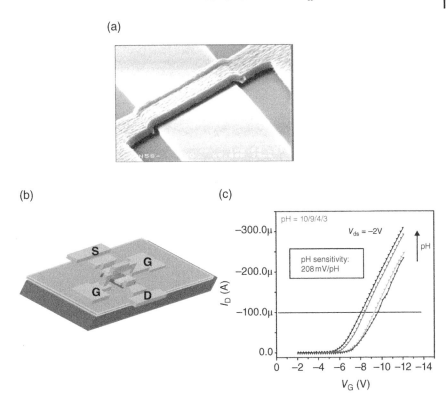

Figure 2.9 MEB image of a suspended gate structure (a), scheme of the SGFET (b), and response to pH (c). *Source:* From Bendriaa et al. [46], reprinted with permission.

can be significantly higher than the limitation of Nernst. Different technologies involving SOI layers [47], thin film technologies [48], organic materials [49], or nanostructures [50] are proposed. In all cases, the detection method is the same, but the enhancement of the sensitivity depends on the technological parameters. Examples of structures are given in Figure 2.10.

The sensitivity for pH detection can be as high as 948 mV/pH with nanostructures [50], or even reach 2.25 V/pH [47].

2.3.2.4   Water Gating FET or Electrolyte Gated FET

Another technology based on FET was recently developed and is called the water gating FET. Its principle is, as shown in Figure 2.11, based on an electrolyte present between the channel and the suspended gate of the transistor. This device uses the electrical double layer formed in the liquid as the gate insulator and operates at low voltage causing no electrochemical reactions [52, 53].

(a)

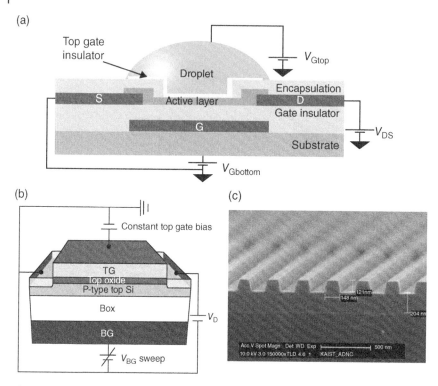

Figure 2.10 Examples of dual gate structures. *Source:* From Le Bihan et al. [48] (a), Lim et al. [50] (b, c), reprinted with permission.

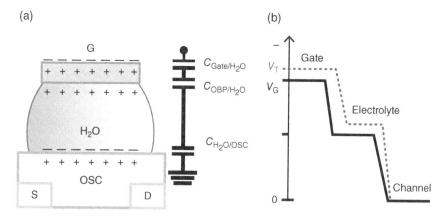

Figure 2.11 Water gating mechanism with capacitance modeling (a) and electrical potential variations (b). *Source:* From Mulla et al. [51] (open access).

The most important advantage of this technology is that the detection of a charged bioelement can occur at a distance higher than the Debye length (see Section 2.3.3.1) [54] and is then compatible with bigger bioelements as specific proteins.

2.3.2.5 Other FETs

Emerging technologies are also used for chemical or biological detection using FET-like structures. It is the case of many organic materials, including 2D materials such as MoS_2, graphene, and other nanomaterials as further described.

Organic FETs have huge interest for chemical or biological sensing. The first one is that many different organic materials exist and a precise synthesis can help to increase the sensing performances: not only the sensitivity, but also the selectivity. It is also possible to directly integrate a recognition element in the material structure for specific interactions [55].

2D materials also offer interesting properties for chemical and biological layers, and can help to reach very low detection limits, as for example a change of 0.025 in the pH value. Graphene layers can be used as active layer [56] or MoS_2 [57].

2.3.3 BioFETs

2.3.3.1 General Considerations

The first ISFETs modified with enzyme were proposed for urea by Danielsson et al. [58] and for the detection of penicillin by Caras and Janata [59]. The ISFET is called then Enzyme FET (ENFET). ISFET can be sensitive to different organic molecules by immobilizing a suitable enzyme layer or specific chemical bounds on its surface. Then, several targets can be detected for different biological analysis as DNA, enzymes, proteins, and even cells. The major limitation for those sensors is the Debye Length (Figure 2.12). Indeed, the screening of charges, especially in electrolytes, cannot be neglected compared to the size of the biological elements, then drastically decreasing the sensitivity of charges detection.

2.3.3.2 DNA BioFET

DNA contains negative charges that can be detected when linked to the ISFET surface. The specificity is obtained by previously linking complementary single-stranded oligonucleotides on the surface. Many examples can be found in literatures either with classical ISFET [60], extended gate FET [61], or SGFET [62] (Figure 2.13). Polysilicon thin film transistors (TFTs) were also used for these kind of applications [63].

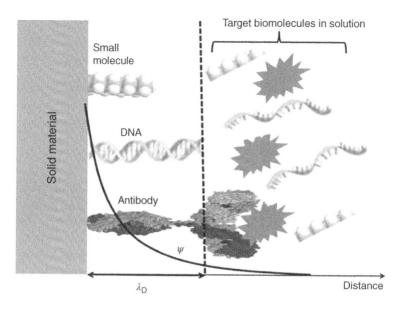

Figure 2.12 Different bindings of biomolecules inside and outside the electrical double layer. Only the changes inside the electrical double layer are detected (λ_D, Debye length). *Source:* From Huang et al. [19], reprinted with permission.

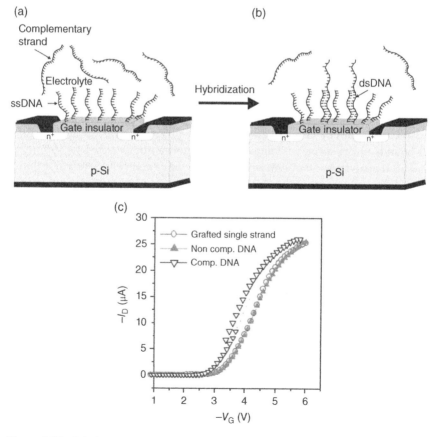

Figure 2.13 Principle of DNA recognition (a), hybridization (b), and example of detection (c). *Source:* From Bendriaa et al. [62], reprinted with permission.

2.3.3.3 Protein BioFET

Protein sensors were also fabricated with FET devices. The biorecognition element is usually an antibody previously bounded to the surface by specific chemical functionalization. These kind of sensors can be achieved with ISFET technologies, Dual-Gate, Extended-Gate, or Water-Gated transistors as shown in Figure 2.14.

2.3.3.4 Cells

Cells contain different types of chemicals, ions, proteins, and other organic molecules. So FET-based sensors can give much information about cell media, extracellular exchange, extracellular pH, concentration of ions as potassium and sodium, oxygen rate, CO_2 production, and so on. Even in the case of specific cells as neurons (Figure 2.15), FET sensors can be used to measure extracellular and intracellular potentials.

2.4 Nano-Field Effect Transistors

The More than Moore approach allows electronics to become smaller, more powerful, and more efficient, and in this way, several innovative devices emerge as a new promising structure with the integration of new nano-objects such as nanowires (NWs), carbon nanotubes (CNTs), and nanomaterials into CMOS chips. Because of their high surface-to-volume ratio, nanostructures are potential functional probes for detecting charged species, their dimensions being comparable with those of chemical and biological species. Nano field-effect transistors (nano-FETs) as biochemical nanosensors are promising for label-free, real-time, and sensitive detection of biomolecules. These technologies may be associated to specific sensitive layers to address selective chemical detection. The miniaturization of biological sensors has several advantages: reduced consumption of samples and reagents, increased sensitivity, and cost-effective disposable chips due to mass production [66]. These sensors have shown a great interest in the numerous studies that have monitored biological events such as various kinds of biorecognition materials for biological analysis such as DNA, proteins, or viruses.

2.4.1 Fabrication of Nano-Devices

2.4.1.1 Silicon Nano-Devices

The growth of silicon nanowires (SiNWs) can be achieved in a bottom-up approach or in a top-down technique. The bottom-up approach is based on self-assembly growth mechanism [67, 68] and the most popular strategy is the Vapor–Liquid–Solid (VLS) growth technique that uses metallic nanoparticles [69] (Figure 2.16). The VLS method enables to achieve a large area production of

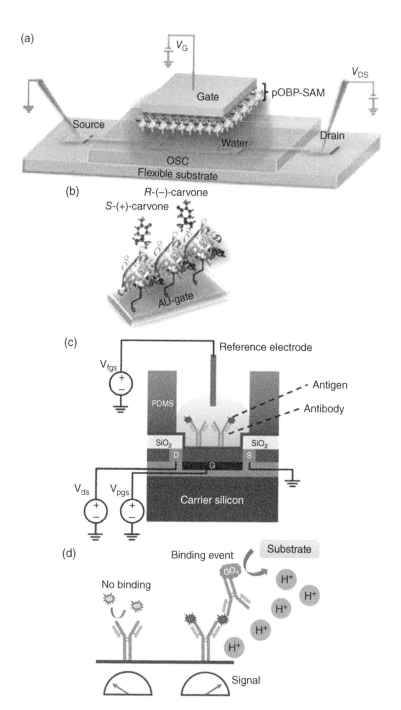

Figure 2.14 Example of water-gated bio-organic transistor (a, b) functionalized with odorant protein for olfactive sensor (*Source:* From Mulla et al. [51], open access), and dual-gate (c, d) for DG-ISFET cheap (*Source:* From Juang et al. [64], reprinted with permission).

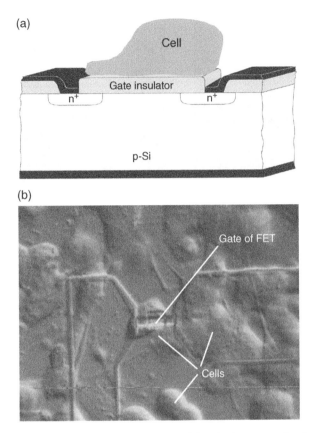

Figure 2.15 Cell bio FET, structure (a) and microscopy of cells on the FET surface (b). *Source:* From Schöning and Poghossian [65], reprinted with permission.

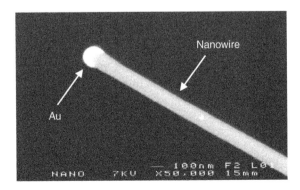

Figure 2.16 Nanowires grown by the VLS process using gold nanoparticle.

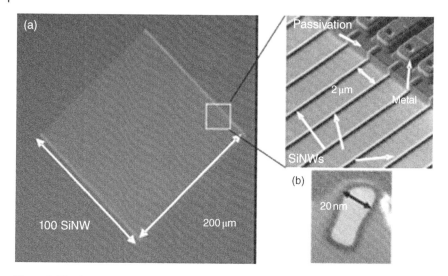

Figure 2.17 Silicon nanowires fabricated by top-down approach (a) image of 100 nanowires, 200 μm long and zoom shows nanowires. (b) TEM image of nanowire showing rectangular cross-section. *Source:* From Agarwal et al. [73], reprinted with permission.

silicon nanowires avoiding using high-cost advanced lithographic tools and allows high quality monocrystalline nanowires with well-controlled composition and electronic properties.

The top-down approach is based on the downscaling technologies, starting from bulk materials and scales down the patterned areas using various techniques [70–72]; but the most common approach is electron beam lithography to form silicon nanowires on silicon-on-insulator (SOI) substrates (Figure 2.17) [73–75].

2.4.1.2 Carbon Nanotubes Nano-Devices

CNTs can be synthesized as single-walled carbon nanotubes (SWNTs) composed of a single layer of carbon atoms, using hollow cylinders of graphene (Figure 2.18a). They benefit to very high aspect ratios, with diameters in the order of 1 nm and several micrometers lengths. If there are several graphene layers, CNTs are classified as multi-walled carbon nanotubes (MWNTs). Single-wall CNTs can be assembled to produce carbon nanotube field effect transistors (CNTFETs) (Figure 2.18b).

2.4.2 Detection of Biochemical Particles by Nanostructures-Based FET

The functionalization of nanostructures is based on the well-known chemical modification of silicon oxide surfaces obtained for planar chemical and biological

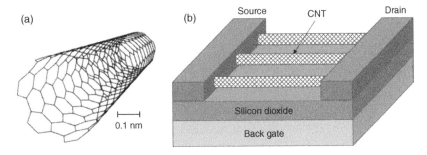

Figure 2.18 (a) Single-walled carbon nanotube made of a rolled-up cylinder of graphene. *Source:* From Saethera et al. [76], reprinted with permission. (b) Schematic multi-channeled carbon nanotube field effect transistor (CNTFET).

sensors. Research efforts over the last decade have led to significant advances in nano-FET biosensor technology, in which specific detection is achieved by connecting a recognition group to the surface of the nanowire. An advantage of the nano-FET compared to a planar-FET is that it can benefit from better electrostatic control of the channel. When the device with functionalized nanowires is exposed to a solution containing macromolecules with net positive or negative charge, it leads to a change in the charge density at the surface, the NW is thus depleted/accumulated by an equal amount of charges, and thus a variation of the conductance. The studies demonstrate interesting capabilities of high sensitive nanostructures-based sensors for the detection at low concentrations of proteins, nucleic acids, and viruses in solution.

2.4.2.1 SiNW pH Sensor

The first pH sensor based on VLS SiNW FET was proposed by the group of Lieber [77]. The amino and silanol act as receptors for hydrogen ions, based on protonation/deprotonation reactions, thereby changing the net nanowire surface charge. Modified p-type Si nanowire devices have shown progressive increases in conductance, as the pH of the solution has gradually increased from 2 to 9. SiNW FETs fabricated using SOI substrates show change of the transistor conductance on a large pH interval [3–10.5], as seen in Figure 2.19 [78].

A lot of authors have also used the interesting properties of SINW FET for pH detection [74, 79]. According to Go et al., modern variants of ISFETs based on silicon nanowires offer a novel form factor to amplify the Nernst signal. While the intrinsic Nernst response of an electrolyte/site-binding interface is 59 mV/pH, they show that the extrinsic sensor response can be increased drastically beyond the Nernst limit [80].

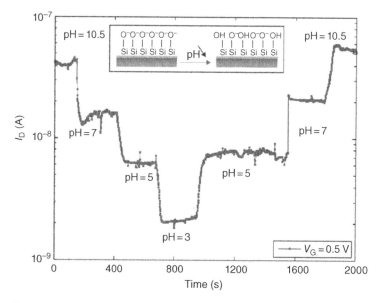

Figure 2.19 Reversible pH monitoring on a large range of pH values: drain current recorded versus time for pH values from 10.5 to 3. *Source:* From Lehoucq et al. [78], reprinted with permission.

2.4.2.2 DNA Detection Using SiNW-Based Sensor

DNA detection technology has been developed rapidly because of its extensive use in the areas of clinical diagnosis, bioengineering, environmental monitoring, and food. Negative charges carried by the phosphate groups of DNA bound onto NW surface (acting as chemical gates) play a significant role in carrier transport in the nanowire channel. According to Streifer et al., the biomolecular recognition properties of the nanowires were tested via hybridization with fluorescently tagged complementary and noncomplementary DNA oligonucleotides, showing good selectivity and reversibility [81]. C.M. Lieber and colleagues investigated a high-sensitive DNA sensor based on nanowires for the detection of single-stranded DNA down to 10 fM level [77]. Wenga et al. have performed probe immobilization by functionalization of poly-SiNW [82] with a procedure described in Figure 2.20. Hybridization phenomenon is detected by electrical measurement at concentrations as low as 1 fM.

Other results have been presented by many authors: for example specific detection of single-stranded oligonucleotides using silicon nanowires [79, 83, 84], and detection limit down to 10 fM obtained for hybridization of peptide nucleic acid (PNA)–DNA (Figure 2.21) [85]. A SiNW biosensor platform based on PNA–DNA hybridization for highly sensitive and rapid detection of Dengue virus has been demonstrated with a detection below 10 fM concentration within 30 minutes [86].

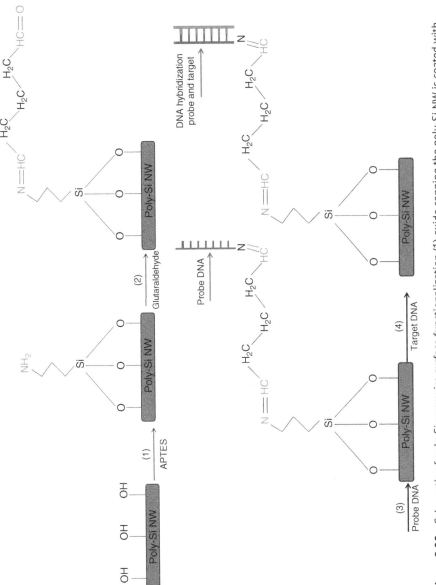

Figure 2.20 Schematic of poly-Si nanowire surface functionalization (1) oxide capping the poly-Si NW is coated with APTES, (2) glutaraldehyde linked to the amino groups, (3) the DNA probe reacts with the aldehyde groups, (4) complementary DNA target is hybridized with the DNA probe. *Source:* From Wenga et al. [82], reprinted with permission.

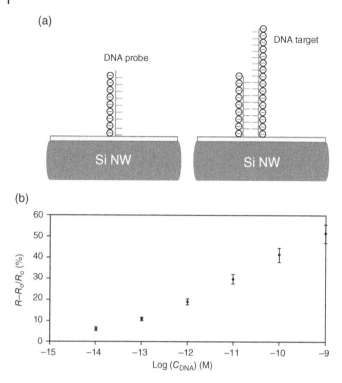

Figure 2.21 (a) Schematic probe binding onto functionalized surface and hybridization with complementary target. (b) Response of the PNA-functionalized SiNWs to target DNA of different concentrations. *Source:* From Zhang et al. [85], reprinted with permission.

CNT-based sensors have also been developed to detect DNA [87]. Some works have reported the development of a nanosensor array based on CNTFETs for DNA hybridization detection [88], with a detection limit of 6.8 fM.

2.4.2.3 Protein Detection

SiNWs have been also used in the detection of various protein molecules and recent studies have indicated that they may be used for sensitive and selective real-time detection of cancer marker proteins. For example, biotin-functionalized SiNWs have been used for the label-free detection of streptavidin [23], schemed in Figure 2.22a. When solutions of streptavidin protein are delivered to nanowire sensor devices, the conductance increases rapidly to a constant value. Kong et al. demonstrate the electrical detection of cardiac troponin (cTnI), a highly sensitive and selective biomarker of acute myocardial infarction using SiNW-based FETs down to $0.092\,\mathrm{ng\,ml^{-1}}$ (Figure 2.22b).

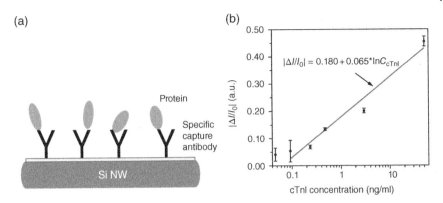

(a)

(b)

$|\Delta I/I_0| = 0.180 + 0.065 * \ln C_{cTnI}$

Figure 2.22 (a) Schematic illustrating antigen target detected with specific antibody and (b) normalized Ids change versus the logarithm of cTnI protein concentrations with SiNW based FET biosensor. *Source:* From Kong et al. [89], reprinted with permission.

Some authors have developed measurements with these SiNWs devices to detect cancer markers using multiplexed real-time monitoring of protein markers [90–93] or for a Lyme disease antigen at concentrations as low as $1\,ng\,ml^{-1}$ [94]. In another example, Zheng et al. used nanowire arrays allowing highly selective and sensitive multiplexed detection of prostate-specific antigen PSA [95].

2.4.2.4 Detection of Bacteria and Viruses

Many other techniques have been reported for pathogen detection. Viruses, expressing specific surface proteins, can be recognized via antibody–antigen interactions [96]. The detection of a viral particle was demonstrated by Patolsky et al. [97] using a SiNW FET functionalized with an *influenza A* surface protein-specific

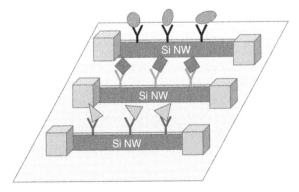

Figure 2.23 Schematic illustrating of selective multiplexed detection: different targets detected with specific antibodies.

antibody, *antihemagglutinin*. The work reported by Wang et al. [98] concerns Gram-negative *Escherichia coli* bacteria used to perform antibacterial activity assays, using the grafting of antibacterial materials on the surface of nanostructured silicon nanowire arrays. A biosensor for the selective determination of *Salmonella Infantis*, based on an FET in which a network of SWCNTs acts as the conductor channel, has been reported using *anti-Salmonella* antibodies with at least $100\,cfu\,ml^{-1}$ in one hour [99].

Another attractive point for nanomaterial FETs is their p`otential integration into electrically addressable sensor arrays. Modified nanowire field effect sensors with specific surface receptors can be integrated on detection platforms selective to multiplexed detection for a wide range of biological and chemical species (Figure 2.23) [100].

This could have a significant impact, for example, on the diagnosis of the disease and genetic screening, and thus constitute powerful new research tools in many areas of biology.

References

1 Capone, S., Zuppa, M., Presicce, D.S. et al. (2008). Metal oxide gas sensor array for the detection of diesel fuel in engine oil. *Sensors and Actuators B: Chemical B* 131: 125–133.

2 Dufour, N. (2013). Conception et réalisation d'un multicapteur de gaz intégré à base de plateformes chauffantes sur silicium et de couches sensibles à oxyde métallique pour le contrôle de la qualité de l'air habitacle. PhD dissertation. Université Paul Sabatier-Toulouse III.

3 Fleischer, M. (2008). Advances in application potential of adsorptive-type solid state gas sensors: high-temperature semiconducting oxides and ambient temperature GasFET devices. *Measurement Science and Technology* 19: 042001.

4 Korotcenkov, G. (2005). Gas response control through structural and chemical modification of metal oxide films: state of the art and approaches. *Sensors and Actuators B: Chemical* 107: 209–232.

5 Korotcenkov, G. (2007). Metal oxides for solid-state gas sensors: what determines our choice? *Materials Science and Engineering B* 139: 1–23.

6 Meixner, H., Gerblinger, H.J., Lampe, U. et al. (1995). Thin-film gas sensors based on semiconducting metal oxides. *Sensors and Actuators B: Chemical* 23: 119–125.

7 Ting Ng, K., Boussaid, F., and Bermak, A. (2011). A CMOS single-chip gas recognition circuit for metal oxide gas sensor arrays. *IEEE Transactions on Circuits and Systems I: Regular Papers* 58: 1569–1580.

8 Wang, C., Yin, L., Zhang, L. et al. (2010). Metal oxide gas sensors: sensitivity and influencing factors. *Sensors* 10 (3): 2088–2106.

9 Bai, H. and Shi, G. (2007). Gas sensors based on conducting polymers. *Sensors* 7 (3): 267–307.

10 Chen, X., Yuan, C., Wong, C.K.Y. et al. (2012b). Molecular modeling of temperature dependence of solubility parameters for amorphous polymers. *Journal of Molecular Modeling* 18 (6): 2333–2341.

11 Chen, X.P., Yuan, C.A., Wong, C.K. et al. (2011). Validation of force fields in predicting the physical and thermophysical properties of emeraldine base polyaniline. *Molecular Simulation* 37 (12): 990–996.

12 Diaz, A.F., Rubinson, J.F., and Mark, H.B. Jr. (1988). Electrochemistry and electrode applications of electroactive/conductive polymers. In: *Electronic Applications*, 113–139. Springer.

13 Bartlett, P.N. and Ling-Chung, S.K. (1989). Conducting polymer gas sensors. Part III: results for four different polymers and five different vapours. *Sensors and Actuators* 20: 287–292.

14 Lundstrom, I., Shlvaraman, M.S., Svensson, C. et al. (1975). A hydrogen sensitive MOS field-effect transistor. *Applied Physics Letters* 26: 55–57.

15 Leu, M., Doll, T., Flietner, B. et al. (1994). Evaluation of gas mixtures with different sensitive layers incorporated in hybrid FET structures. *Sensors and Actuators B: Chemical* 18–19: 678–681.

16 Pohle, R., Simon, E., Fleisher, M. et al. (2003). Realization of a new sensor concept: improved CCFET and SGFET type gas sensors in Hybrid Flip-Chip technology. In: *Proceedings of the 12th International Conference on Solid-State Sensors, Actuators and Microsystems, TRANSDUCERS, Boston, MA (8–12 June 2003)*, vol. 1, 135–138. IEEE.

17 Shi, W., Yu, X.G., Zheng, Y.F. et al. (2016). DNA based chemical sensor for the detection of nitrogen dioxide enabled by organic field-effect transistor. *Sensors and Actuators B: Chemical* 222: 1003–1011.

18 Han, S.J., Huang, W., Shi, W. et al. (2014). Performance improvement of organic field-effect transistor ammonia gas sensor using ZnO/PMMA hybrid as dielectric layer. *Sensors and Actuators B: Chemical* 203: 9.

19 Huang, W., Diallo, A.K., Dailey, J.L. et al. (2015). Electrochemical processes and mechanistic aspects of field-effect sensors for biomolecules. *Journal of Materials Chemistry C* 3: 6445.

20 Hu, W., Liu, Y., Xu, Y. et al. (2000). The gas sensitivity of a metal-insulator-semiconductor field-effect-transistor based on Langmuir–Blodgett films of a new asymmetrically substituted phthalocyanine. *Thin Solid Films* 360: 256–260.

21 Muzikante, I., Parra, V., Dobulans, R. et al. (2007). A novel gas sensor transducer based on phthalocyanine heterojunction devices. *Sensors* 7: 2984.

22 Cao, A., Sudhölter, E., and de Smet, L. (2013). Silicon nanowire-based devices for gas-phase sensing. *Sensors* 14 (1): 245–271.

23 Cui, Y., Wei, Q., Park, H. et al. (2001b). Nanowire nanosensors for highly sensitive and selective detection of biological and chemical species. *Science* 293 (5533): 1289–1292.

24 Engel, Y., Elnathan, R., Pevzner, A. et al. (2010). Supersensitive detection of explosives by silicon nanowire arrays. *Angewandte Chemie, International Edition* 49 (38): 6830–6835.

25 Kolmakov, A. and Moskovits, M. (2004). Chemical sensing and catalysis by one-dimensional metal-oxide nanostructures. *Annual Review of Materials Research* 34 (1): 151–180.

26 Liao, L., Zhang, Z., Yan, B. et al. (2009). Multifunctional CuO nanowire devices: p-type field effect transistors and CO gas sensors. *Nanotechnology* 20 (8): 085203.

27 Zhang, D., Liu, Z., Li, C. et al. (2004). Detection of NO_2 down to ppb levels using individual and multiple In_2O_3 nanowire devices. *Nano Letters* 4 (10): 1919–1924.

28 McAlpine, M.C., Ahmad, H., Wang, D. et al. (2007). Highly ordered nanowire arrays on plastic substrates for ultrasensitive flexible chemical sensors. *Nature Materials* 6 (5): 379–384.

29 McAlpine, M.C., Friedman, R.S., Jin, S. et al. (2003). High performance nanowire electronics and photonics on glass and plastic substrates. *Nano Letters* 3 (11): 1531–1535.

30 Fan, Z. and Lu, J.G. (2006). Gate-refreshable nanowire chemical sensors. *Applied Physics Letters* 86: 123510–123512.

31 Li, C., Zhang, D., Liu, X. et al. (2003). In_2O_3 nanowires as chemical sensors. *Applied Physics Letters* 82: 1613–1616.

32 Bergveld, P. (1970). Development of an ion-sensitive solid-state device for neurophysiological measurements. *IEEE Transactions on Biomedical Engineering* BME-17: 70–71.

33 Bergveld, P. (2003). Thirty years of ISFETOLOGY: what happened in the past 30 years and what may happen in the next 30 years. *Sensors and Actuators B: Chemical* 88: 1–20.

34 Van Hal, R.E.G., Eijkel, J.C.T., and Bergveld, P. (1995). A novel description of ISFET sensitivity with the buffer capacity and double-layer capacitance as key parameters. *Sensors and Actuators B: Chemical* 24: 201–205.

35 Yates, D.E., Levine, S., and Healy, T.W. (1974). Site-binding model of the electrical double layer at the oxide/water interface. *Journal of the Chemical Society, Faraday Transactions* 70: 1807–1818.

36 Bousse, L., De Rooij, N.F., and Bergveld, P. (1983). Operation of chemically sensitive field-effect sensors as a function of the insulator-electrolyte interface. *IEEE Transactions on Electron Devices* 30 (10): 1263–1270.

37 Baccar, Z.M., Jaffrezic-Renault, N., Martelet, C. et al. (1996). Sodium microsensors based on ISFET/REFET prepared through an ion-implantation process fully compatible with a standard silicon technology. *Sensors and Actuators, B: Chemical* 32: 101–105.

38 Da Silva Rodrigues, B., De Sagazan, O., Crand, S. et al. (2009). Sensitive continuous monitoring of PH thanks to matrix of several suspended gate field effect transistors. *ECS Transactions* 23 (1): 203–209.

39 Jimenez-Jorquera, C., Orozco, J., and Baldi, A. (2010). ISFET based microsensors for environmental monitoring. *Sensors* 10: 61–83.

40 Joly, M., Mazenq, L., Marlet, M. et al. (2017). Multimodal probe based on ISFET electrochemical microsensors for in-situ monitoring of soil nutrients in agriculture. *Proceedings* 1: 420.

41 Sasipongpana, S., Rayanasukha, Y., Prichanont, S. et al. (2017). Extended–gate field effect transistor (EGFET) for carbaryl pesticide detection based on enzyme inhibition assay. *Materials Today: Proceedings* 4: 6458–6465.

42 Batista, P.D. and Mulato, M. (2005). ZnO extended-gate field-effect transistors as pH sensors. *Applied Physics Letters* 87: 143508.

43 Chi, L.-L., Chou, J.-C., Chung, W.-Y. et al. (2000). Study on extended gate field effect transistor with tin oxide sensing membrane. *Materials Chemistry and Physics* 63: 19–23.

44 Van der Spiegel, J., Lauks, I., Chan, P., and Babic, D. (1983). The extended gate chemically sensitive field effect transistor as multi-species microprobe. *Sensors and Actuators* 4: 291–298.

45 Park, J.-K. and Cho, W.-J. (2012). Development of high-performance fully depleted silicon-on-insulator based extended-gate field-effect transistor using the parasitic bipolar junction transistor effect. *Applied Physics Letters* 101: 133703.

46 Bendriaa, F., Le Bihan, F., Salaün, A.C. et al. (2006). Study of mechanical maintain of suspended bridge devices used as pH sensor. *Journal of Non-Crystalline Solids* 352 (9–20): 1246–1249.

47 Jang, H.-J. and Cho, W.-J. (2014). Performance enhancement of capacitive-coupling dual-gate ion-sensitive field-effect transistor in ultra-thin-body. *Scientific Reports* 4: 5284.

48 Le Bihan, F., Donero, L., Le Borgne, B. et al. (2018). Dual-gate TFT for chemical detection. *ECS Transactions, Thin Film Transistor Technologies 14 (TFTT 14)* 86 (11): 169–176.

49 Spijkman, M.-J., Brondijk, J.J., Geuns, T.C.T. et al. (2010). Dual-gate organic field-effect transistors as potentiometric sensors in aqueous solution. *Advanced Functional Materials* 20: 898–905.

50 Lim, C.-M., Lee, I.-K., Lee, K.-J. et al. (2017). Improved sensing characteristics of dual-gate transistor sensor using silicon nanowire arrays defined by nanoimprint lithography. *Science and Technology of Advanced Materials* 18 (1): 17–25.

51 Mulla, M.Y., Tuccori, E., Magliulo, M. et al. (2015). Capacitance-modulated transistor detects odorant binding protein chiral interactions. *Nature Communications* 6: 6010.

52 Sonmez, B., Ertop, O., and Mutlu, S. (2017). Modelling and realization of a water-gated field effect transistor (WG-FET) using 16-nm-thick, mono-Si film. *Scientific Reports* 7: 12190.

53 Wang, D., Noël, V., and Piro, B. (2016). Electrolytic gated organic field-effect transistors for application in biosensors – a review. *Electronics* 5: 9.

54 Palazzo, G., De Tullio, D., Magliulo, M. et al. (2014). Detection beyond Debye's length with an electrolyte-gated organic field-effect transistor. *Advanced Materials* 27: 911–916.

55 Mabeck, J.T. and Malliaras, G.G. (2006). Chemical and biological sensors based on organic thin-film transistors. *Analytical and Bioanalytical Chemistry* 384: 343–353.

56 Ohno, Y., Maehashi, K., and Matsumoto, K. (2010). Chemical and biological sensing applications based on graphene field-effect transistors. *Biosensors and Bioelectronics* 26: 1727–1730.

57 Sarkar, D., Liu, W., Xie, X. et al. (2014). MoS_2 field-effect transistor for next-generation label-free biosensors. *ACS Nano* 8 (4): 3992–4003.

58 Danielsson, B., Lundström, I., Mosbach, K. et al. (1979). A new enzyme-transducer combination-the enzyme transistor. *Analytical Letters* 12: 1189–1195.

59 Caras, S. and Janata, J. (1980). Field effect transistor sensitive to penicillin. *Analytical Chemistry* 52: 1935–1937.

60 Ingebrandt, S., Han, Y., Nakamura, F. et al. (2007). Label-free detection of single nucleotide polymorphisms utilizing the differential transfer function of field-effect transistors. *Biosensors and Bioelectronics* 22: 2834–2840.

61 Kamahori, M., Ishige, Y., and Shimoda, M. (2007). DNA detection by an extended-gate FET sensor with a high-frequency voltage superimposed onto a reference electrode. *Analytical Sciences* 23: 75.

62 Bendriaa, F., Le-Bihan, F., Salaün, A.C. et al. (2005). DNA detection by suspended gate polysilicon thin film transistor. *Proceeding IEEE Sensors*, Irvine, California (31 October–3 November 2005).

63 Estrela, P., Stewart, A.G., Yan, F. et al. (2005). Field effect detection of biomolecular interactions. *Electrochimica Acta* 50: 4995–5000.

64 Juang, D.S., Lin, C.-H., Huo, Y.-R. et al. (2018). Proton-ELISA: electrochemical immunoassay on a dual-gated ISFET array. *Biosensors and Bioelectronics* 117: 175–182.

65 Schöning, M.J. and Poghossian, A. (2002). Recent advances in biologically sensitive field-effect transistors (BioFETs). *Analyst* 127: 1137–1151.

66 Furuberg, L., Mielnik, M., Gulliksen, A. et al. (2008). RNA amplification chip with parallel microchannels and droplet positioning using capillary valves. *Microsystem Technologies* 14: 673–681.

67 Lee, E.K., Choi, B.L., Park, Y.D. et al. (2008). Device fabrication with solid–liquid–solid grown silicon nanowires. *Nanotechnology* 19: 185701.

68 Rabin, O., Herz, P.R., Lin, Y.-M. et al. (2003). Formation of thick porous anodic alumina films and nanowire arrays on silicon wafers and glass. *Advanced Functional Materials* 13 (8): 631–638.

69 Cui, Y., Lauhon, L.J., Gudiksen, M.S. et al. (2001a). Diameter-controlled synthesis of single-crystal silicon nanowires. *Applied Physics Letters* 78 (15): 2214–2216.

70 Demami, F., Pichon, L., Rogel, R. et al. (2009). Fabrication of polycrystalline silicon nanowires using conventional UV lithography. *Materials Science and Engineering* 6: 012014.

71 Hsiao, C.Y., Lin, C.H., Hung, C.H. et al. (2009). Novel poly-silicon nanowire field effect transistor for biosensing application. *Biosensors and Bioelectronics* 24: 1223–1229.

72 Lin, C.-H., Hung, C.-H., Hsiao, C.-Y. et al. (2009). Poly-silicon nanowire field-effect transistor for ultrasensitive and label-free detection of pathogenic avian influenza DNA. *Biosensors and Bioelectronics* 24: 3019–3024.

73 Agarwal, A., Buddharaju, K., Lao, I.K. et al. (2008). Silicon nanowire sensor array using top–down CMOS technology. *Sensors and Actuators A: Physical* 145–146: 207–213.

74 Park, I., Zhiyong, L., Pisano, A.P. et al. (2010). Top-down fabricated silicon nanowire sensors for real-time chemical detection. *Nanotechnology* 21: 015501.

75 Sacchetto, D., Ben-Jamaa, M.H., De Micheli, G. et al. (2009). Si nanowire effect transistors for low current and temperature sensing. *Proceedings ESSDERC*, pp. 245–248. Piscataway, NJ: IEEE.

76 Saethera, E., Franklandb, S.J.V., and Pipesc, R.B. (2003). Transverse mechanical properties of single-walled carbon nanotube crystals. Part I: determination of elastic moduli. *Composites Science and Technology* 63: 1543–1550.

77 Patolsky, F. and Lieber, C.M. (2005). Nanowire nanosensors. *Materials Today* 8: 22–26.

78 Lehoucq, G., Bondavalli, P., Xavier, S. et al. (2012). Highly sensitive pH measurements using a transistor composed of a large, array of parallel silicon nanowires. *Sensors and Actuators B: Chemical* 171–172: 127–134.

79 Chen, M.-C., Chen, H.-Y., Lin, C.-Y. et al. (2012a). A CMOS-compatible poly-Si nanowire device with hybrid sensor/memory characteristics for system-on-chip applications. *Sensors* 12: 3952–3963.

80 Go, J., Nair, P.R., Reddy, B. et al. (2012). Coupled heterogeneous nanowire-nanoplate planar transistor sensors for giant (>10 V/pH) Nernst response. *ACS Nano* 6 (7): 5972–5979.

81 Streifer, J.A., Kim, H., Nichols, B.M. et al. (2005). Covalent functionalization and biomolecular recognition properties of DNA-modified silicon nanowires. *Nanotechnology* 16: 1868–1873.

82 Wenga, G., Jacques, E., Salaun, A.-C. et al. (2013). Step-gate polysilicon nanowires field effect transistor compatible with CMOS technology for label-free DNA biosensor. *Biosensors and Bioelectronics* 40: 141–146.

83 Maki, W.C., Mishra, N.N., Cameron, E.G. et al. (2008). Nanowire-transistor based ultra-sensitive DNA methylation detection. *Biosensors and Bioelectronics* 23: 780–787.

84 Singh, S., Zack, J., Kumar, D. et al. (2010). DNA hybridization on silicon nanowires. *Thin Solid Films* 519: 1151–1155.

85 Zhang, G.-J., Chua, J.H., Chee, R.-E. et al. (2008). Highly sensitive measurements of PNA-DNA hybridization using oxide-etched silicon nanowire biosensors. *Biosensors and Bioelectronics* 23: 1701–1707.

86 Zhang, G.-J., Zhang, L., Huang, M.J. et al. (2010). Silicon nanowire biosensor for highly sensitive and rapid detection of Dengue virus. *Sensors and Actuators B: Chemical* 146: 138–144.

87 Khamis, S.M., Jones, R.A., Johnson, A.T.C. et al. (2012). DNA-decorated carbon nanotube-based FETs as ultrasensitive chemical sensors: discrimination of homologues, structural isomers, and optical isomers. *AIP Advances* 2: 022110.

88 Kerman, K., Morita, Y., Takamura, Y. et al. (2005). Peptide nucleic acid–modified carbon nanotube field-effect transistor for ultra-sensitive real-time detection of DNA hybridization. *Nanobiotechnology* 1: 65–70.

89 Kong, T., Su, R., Zhang, B. et al. (2012). CMOS-compatible, label-free silicon-nanowire biosensors to detect cardiac troponin I for acute myocardial infarction diagnosis. *Biosensors and Bioelectronics* 34: 267–272.

90 Ginet, P., Akiyama, S., Takama, N. et al. (2011). CMOS-compatible fabrication of top-gated field-effect transistor silicon nanowire-based biosensors. *Journal of Micromechanics and Microengineering* 21: 065008.

91 Hakim, M.M., Lombardini, M., Sun, K. et al. (2012). Thin film polycrystalline silicon nanowire biosensors. *Nano Letters* 12: 1868–1872.

92 Kim, A., Ah, C.S., Yu, H.Y. et al. (2007). Ultrasensitive, label-free, and real-time immunodetection using silicon field-effect transistors. *Applied Physics Letters* 91: 103901.

93 Kulkarni, A., Xu, Y., Ahn, C. et al. (2012). The label free DNA sensor using a silicon nanowire array. *Journal of Biotechnology* 160: 91–96.

94 Lerner, M.B., Dailey, J., Goldsmith, B.R. et al. (2013). Detecting Lyme disease using antibody-functionalized single-walled carbon nanotube transistors. *Biosensors and Bioelectronics* 45: 163–167.

95 Zheng, G., Patolsky, F., Cui, Y. et al. (2005). Multiplexed electrical detection of cancer markers with nanowire sensor arrays. *Nature Biotechnology* 23 (10): 1294–1301.

96 Mishra, N.N., Maki, W.C., Cameron, E. et al. (2008). Ultra-sensitive detection of bacterial toxin with silicon nanowire transistor. *Lab on a Chip* 8: 868–871.

97 Patolsky, F., Zheng, G.F., Hayden, O. et al. (2004). Electrical detection of single viruses. *Proceedings of the National Academy of Sciences of the United States of America* 101: 14017.

98 Wang, H., Wang, L., Zhang, P. et al. (2011). High antibacterial efficiency of pDMAEMA modified silicon nanowire arrays. *Colloids and Surfaces B: Biointerfaces* 83: 355–359.

99 Villamizar, R.A., Maroto, A., Rius, F.X. et al. (2008). Fast detection of *Salmonella Infantis* with carbon nanotube field effect transistors. *Biosensors and Bioelectronics* 24: 279–283.

100 He, B., Morrow, T.J., and Keating, C.D. (2008). Nanowire sensors for multiplexed detection of biomolecules. *Current Opinion in Chemical Biology* 12 (5): 522–528.

3

Mammalian Cell-Based Electrochemical Sensor for Label-Free Monitoring of Analytes

Md. Abdul Kafi[1], Mst. Khudishta Aktar[1], and Hadi Heidari[2]

[1] *Department of Microbiology and Hygiene, Bangladesh Agricultural University, Mymensingh, Bangladesh*
[2] *School of Engineering, University of Glasgow, Glasgow, UK*

3.1 Introduction

Cellular electrophysiological state is sensitive to any adverse external stimuli such as changes in nutrient, and presence of toxicants, heavy metals, and infectious agents [1]. Thus, the living cell-based sensor platform is considered as a potential tool for monitoring analytes having adverse effect on it. Hence, the electrical readout signals from a living cell immobilized platform has been employed for rapid, sensitive, and real-time monitoring of analytes [2, 3]. Therefore, designing, fabrication, and demonstration of this electrical readout of cellular electrophysiology have attracted much attention these days.

The selectivity, sensitivity, and specificity of the electrical signals from a cell-based chip largely depends on the conductivity of the carrier platform, appropriate measurement methods, and physiological state of the living cells established on it. Therefore, researchers over the globe are searching for suitable materials offering firm adhesion of cells and developing suitable methods for acquiring and amplifying signals from the sensor [4–6]. Many materials such as Gold (Au), Indium Tin Oxide (ITO), Platinum (Pt), etc. were employed as a platform for holding cell-based electrodes [6–9]. Though all the metals discussed earlier are expensive, a nano scale thin layer of these materials were sputtered on a carrier substrate for establishing electrical communication between cell and recorder devices. Kafi et al. established 50 nm Au-layer on silicon for establishing mammalian cells (Neural, HeLa, HEK-293, etc.) for monitoring environmental analytes [6, 10, 11].

Smart Sensors for Environmental and Medical Applications, First Edition. Edited by Hamida Hallil and Hadi Heidari.
© 2020 The Institute of Electrical and Electronics Engineers, Inc.
Published 2020 by John Wiley & Sons, Inc.

Kim et al. established a neural cell on the ITO glass for establishing cell–based chips [7]. Lupu et al. established a platinum-based electrochemical sensor for monitoring determining analytes [12]. Among these materials, Pt and Au are highly compatible with living cells and hence, found to be suitable for establishing electrical communication between cell and electrode effectively. ITO glass showed poor communication since it was not compatible enough with living cells for establishing a firm attachment [13]. Despite proven compatibility, Pt and Au also faced difficulties in holding cells for a longer period because of weak anchoring [14]. Therefore, many researchers worked at the cell–electrode interface for establishing firm adhesion as well as offering an in vivo-like bio-friendly environment for holding cells for a longer period of time.

Recently, the conformability has appeared as an another promising feature of cell-based sensor platforms focusing their possible integration with wearables/implantable devices for monitoring clinical analytes from closer vicinity of a patient [15, 16]. The state-of-the-art cell-based platform employs a nonconformable rigid platform as discussed in the earlier section and hence is not suitable for the said purpose [2, 6–9]. Recently, conductive bio-polymers, synthetic polymers, and their hybrids have attracted much attention for developing cell-based conformable devices. However, most of them suffer from poor electrical conductivity in communicating between cells and electrodes [17]. To overcome this limitation, many researchers established an additional thin conductive layer on the cured polymers for establishing a stable electrical communication. Still, conductivity of the polymer-based flexible platform faces challenges because of a cracking tendency of the conductive layer established on it. Therefore, researchers are focusing on this issue and developing a noble biopolymer material for wearable cell-based devices.

Integration of appropriate recorder devices is another major task for acquisition and amplification of electrical readout signals from the chip for sensitive monitoring of analytes. The potentiometric oscilloscope devices, spectra-analyzer, and other Computer Aided Devices (CADs) have been utilized for recoding and amplifying these signals [18–20]. Other than device type, the analytical methods of each device have also influenced the sensing performance of a cell-based chip. For instance, differential pulse voltammetric (DPV) analysis showed better performances over the cyclic voltammetric (CV) analysis of neural cell chip during electrochemical monitoring of environmental toxicants [21–23]. Hence, many researchers are focusing on the integration of appropriate detection devices and analytical methods for enhancing the sensing performances of the cell chip. The integration of RFID systems on the chip platform has also attracted much attention because of their necessity of remote monitoring in future [24].

Based on these issues and promises related to cell-based sensing devices, this chapter was designed for discussing the state-of-the-art cell chip design and fabrication, substrate functionalization strategies at the cell–electrode interface, electrochemical characterization of cellular redox, application of cell-specific redox for monitoring analytes, and finally concluded with the future prospects of cell-based sensors.

3.2 State-of-the-Art Cell Chip Design and Fabrication

The fabrication of a cell-based chip employs the establishment of conductive, compatible metals on the carrier substrate as discussed before; affixing a cell chip chamber for holding cell culture medium and electrolytes; and integration of reference and counter electrodes for electrochemical measurement. Many expensive conductive, compatible materials such as Au, Pt, and ITO were established as a nano scale layer on carrier substrate such as Si, glass, and PDMS [2, 6–9]. The establishment of a conductive layer on the carrier substrate is performed straightforward with metallization instruments [25, 26]. But this sometimes involves the establishment of another adhesive layer for establishing the firm attachment on the carrier substrate [26]. For instance, a 50 nm thick Ti layer was established in the Si substrate and then a 150 nm thick Au layer was patterned by DC magneto sputtering (Figure 3.1).

Since past decades cell-based sensing has experienced much improvement in its architectural design from centimeter to millimeter scale for holding several thousands to a single cell. The cell chip technology emerges from the immobilization of living cells on conductive substrate like Au, Pt, and ITO glass for recording cellular redox (Figure 3.2a). During this time, the chip chambers were prepared by

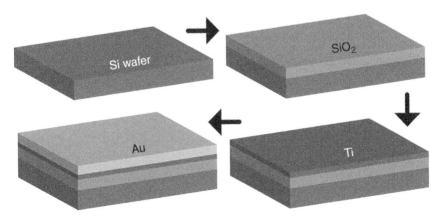

Figure 3.1 Schematic illustration of metal electrode fabrication steps for holding a cell-based chip.

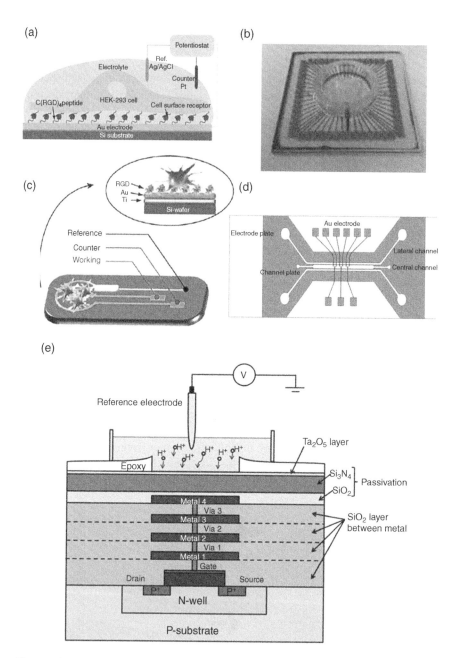

Figure 3.2 State-of-the-art cell chip design: (a) Cell immobilized on Si-based Au electrode with vertically oriented reference and counter electrode. *Source:* Reproduced with permission from Ref. [2]. (b) Microelectrode array with chip chamber for holding cell culture medium. *Source:* Reproduced with permission from Ref. [27]. (c) In-built working, counter, and reference electrodes on a Si-based chip. *Source:* Reproduced with permission from Ref. [21]. (d) Integration of microfluidic channel on a microelectrode array-based cell chip. *Source:* Reproduced with permission from Ref. [28]. (e) An integrated circuit for chip-based analysis. *Source:* Reproduced with permission from Ref. [29]. (f) A handheld high-sensitivity micro-NMR CMOS platform for biological assays. *Source:* Reproduced with permission from Ref. [30].

(f)

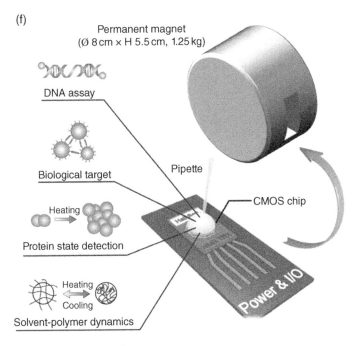

Figure 3.2 (Continued)

affixing plastic chambers of $4 \times 4\,cm^2$, $4 \times 2\,cm^2$, $2 \times 2\,cm^2$, and $1 \times 2\,cm^2$ on those conductive platforms [9, 10]. The cell-seeded chambers were considered as a working electrode while commercially available Ag/AgCl, reference, and Pt, counter electrodes were placed vertically in the chamber (Figure 3.2a). With the advancement of fabrication technologies recently the microelectrode array modified chip were designed by many groups for obtaining cumulated signals from different places of chip surface (Figure 3.2b). With the advancement of microfabrication technologies, recent devices are incorporated with the in-built reference and counter electrode for convenience of handling the chip (Figure 3.2c). A recently developed cell chip was employed in the microfluidics system with inlets for delivering the culture medium and analytes and outlets for expelling the waste materials for real-time monitoring analytes (Figure 3.2d). Therefore, the current state-of-the-art devices are suitable for onsite monitoring of environmental influences. However, the device still requires the incorporation of an RFID tag for monitoring the signal from a distance.

Recently, the ion-sensitive field effect transistor (ISFET) has been proven as an excellent transducer for biosensing applications (Figure 3.2e). The microfabricability of such ISFETs ensures their low unit cost, batch fabrication, and straightforward miniaturization for their use in point-of-care devices. As an advancement in this direction, we can cite the micronuclear magnetic resonance (NMR) system compatible with multi-type biological lab-on-a-chip assays (Figure 3.2f). Integration of such label-free, real-time monitoring systems has attracted much attention and thus many researchers are focusing on the application of such systems for in vivo monitoring of cellular electrophysiological responses in situ [31–33]. The state-of-the-art metal-based platform showed incompatibility for such in vivo applications. The chip conformability is an essential feature for integration with wearable or implantable applications [15, 16]. Therefore, establishment of chips on materials with conformability, biocompatibility, and electrical stability is in recent demand. With this in mind, few groups are designing such electrode platforms on organic or synthetic polymers for achieving their wearability or implantability adequate for integrating them with in vivo living system [16, 34]. However, such platforms are still suffering with poor electrical stability and degradability because of their cracking tendency. Therefore, much attention is required for overcoming these limitations to attain wearability and implantability of cell-based platforms for in situ monitoring of the patient.

3.3 Substrate Functionalization Strategies at the Cell–Electrode Interface

Many cells can be established on the platform without any substrate functionalization while some cells such as neural cells, stem cells, etc. require additional electrode functionalization with appropriate ligand molecules. For instance, poor adhesion was noticed when a neural cell was immobilized on a metal electrode surface [35, 36]. In such occasion, incorporation of ligand molecules (adhesion motifs) was an inevitable phenomena for developing a neural cell chip. Several other cells like embryonic stem cells (ESC), progenitor cells (PC), and induced pluripotent stem cells (IPSC) encountered similar shortcoming when incorporated with artificial surfaces [3, 6, 35]. Researchers all over the globe utilize many biopolymers such as collagen, fibronectin, poly-L-lysine, or short peptide sequences (RGD) as adhesion motifs on the artificial surface for overcoming such challenges [3, 6]. Among these materials, collagen confers numerous focal adhesions while incorporated as a self-assembled (SA) thin layer though it shows incompatibility during electrochemical measurements [37, 38]. The SA layer at the interface of cell–electrode acts as a barrier for electron exchange phenomenon between the cell and the electrode [39]. Therefore, bioengineering of adhesion molecules at the cell–electrode interface has drawn attention. Ruoslahti reported that simple RGD tripeptide enriched portion of collagen molecules are involved in the cell adhesion process through RGD–integrin coupling method [40]. Hence, many groups concentrated on designing and synthesizing RGD peptides for

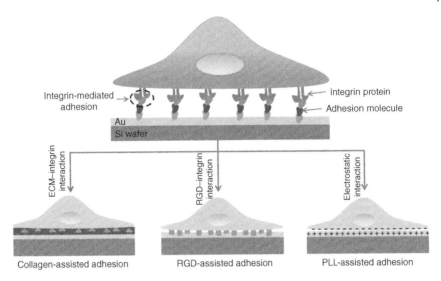

Figure 3.3 Schematic illustrations of various cell–electrode adhesion methods.

utilization as adhesion molecules on artificial surfaces [39–41]. However, the RGD sequences were unable to attach with artificial surfaces in its native form. Therefore, the peptide was engineered with many additional ligands molecules at its terminal region. For instance, cysteine-terminated RGD sequence was found to be suitable for assembling on Au surface through Thiol–Au-coupling method [6, 10]. The self-assembled thiolated RGD peptide functionalized platform showed significant improvement in cell adhesion as well as electrical performance. Following this progress, the thiolated RGD peptide was assembled on the platform in a nano scale pattern aligning with receptor pattern on the cell surface [11, 42, 43]. Such nano scale RGD peptide modified surface showed significant enhancement in cell adhesion, proliferation, and electrochemical performances on the chip. Later on, the special and vertical arrangement of the RGD peptide-modified platforms was also evaluated for enhancing the performance of the functionalized platform (Figure 3.3).

3.4 Electrochemical Characterization of Cellular Redox

The cell, a structural and functional unit of an individual, possesses unique electrophysiology specific to its types and different phases of its growth cycle [10]. This cellular electrophysiological state is sensitive to any endogenous or exogenous

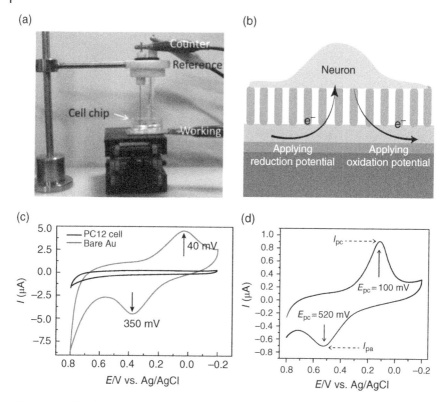

Figure 3.4 Electrochemical measurement of cellular redox: (a) A three-electrode setup for voltammetric measurement of a cell chip. *Source:* Reproduced with permission from Ref. [10]. (b) Electron exchange phenomena at cell–electrode interface. *Source:* Reproduced with permission from Ref. [6]. (c) CV obtained from PC12 cell. *Source:* Reproduced with permission from Ref. [6]. (d) CV obtained from HEK 293. *Source:* Reproduced with permission from Ref. [2].

stimuli that interferes with the homeostatic condition [3]. Recently, the cell line and cell cycle specific electrophysiological redox were employed in the cell-based chip as a sensitive tool for monitoring analytes where the interaction between stimulus and cell immobilized electrode was detected as an electrochemical readout signal using simple electrochemical detection systems (e.g. potentiostat) [3, 10]. Many researchers utilize three electrode configurations, where the cell immobilized platform is used as a working electrode while standard Ag/AgCl, and Pt served as reference and counter electrodes, respectively [2, 9] (Figure 3.4a). Electrochemical analysis of a cell immobilized chip reveals potential peaks of anodic (I_{pa}) and cathodic peak currents (I_{pc}) which are sensitive to any external stimulus: addition of glucose enhances the peaks while exposure of toxicants reduces the peaks [44].

As analytical methods, CV and DPV were utilized for characterizing various cell types (Figure 3.4b). For instance, rat pheochromocytoma (PC12) cells immobilized on Au platform showed quasireversible redox behavior when subjected to CV analysis using potential window -0.2 to $0.8\,V$ at a $100\,mVs^{-1}$ scan rate with an anodic peak at $+40\,mV$ and a cathodic peak at $+350\,mV$ while HeLa cell originated from human endothelium give an anodic peak at $-75\,mV$ and a cathodic peak at $+150\,mV$ (Figure 3.4c and d).

The cell line specificity of redox peaks was confirmed utilizing cells from different origin such as neural cells, kidney cells, hepatic cells, cervix cells, and dermal fibroblast cells. (Figure 3.5a). Another experiment was considered as the anodic peak potential that was obtained from CV technique at a potential window of -0.2 to $0.4\,V$ that was applied to measure DPV from PC12 and HeLa cell line at a scan rate of $100\,mVs^{-1}$ with $50\,mV$ pulse amplitude and $50\,ms$ pulse width [9]. Well-distinguished DPV signals measured from PC12 and Human kidney cell (HEK-293) lines give a peak at 100 and $520\,mV$, respectively, when they are immobilized on Au electrode whereas no such peak was observed from bare Au surface (Figure 3.5b). Furthermore, cell cycle phase specific signals were also detected from synchronized cells at various cell cycle stages (G_1/S, G_2M, and G_0 phase) utilizing many neural and cancerous cell lines (Figure 3.5c). For instance, in an experiment, PC12 cells were synchronized at two major phases of a cell cycle (such as synthesis and mitosis phase) and subjected to DPV measurement. The signals from synchronized cells showed significant differences with nonsynchronized control cells [10]. Even the synchronized phases showed distinctly different DPV signals. Thus, the cell line specificity of the redox signals was proven by both CV and DPV methods.

3.5 Application of Cell-Based Sensor

Cellular redox-based electrochemical monitoring protocol was utilized by many groups all over the globe for monitoring cell viability and cell proliferation assay, environmental toxicity assay, cell cycle measurement, and electroporation of therapeutics or genetic materials [2, 6, 9, 10, 45].

This cell-based electrochemical method was also utilized for label-free, sensitive, real-time monitoring of cell viability. This was achieved by quantification and analysis of peak enhancement or reductions. The cell number-based peak enhancement was reported electrochemically and verified with standard trypan blue exclusion method [10]. The newly developed method was also verified with standard fluorescent-assisted cell sorting method (FACS) (Figure 3.6). The cell numbers obtained with FACS was completely correlated with the corresponding I_{pc} values measured from respective samples and thus confirms the validation of the newly developed method.

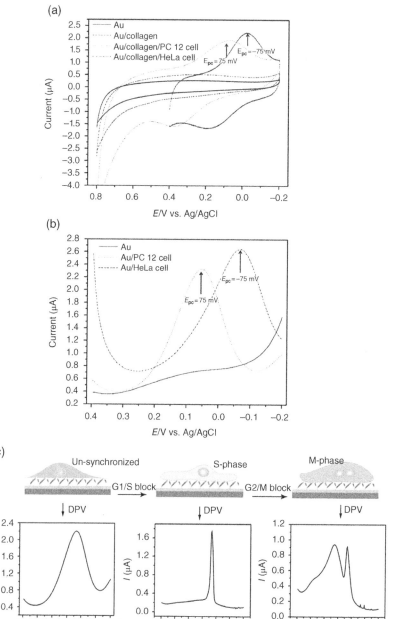

Figure 3.5 Electrochemical measurement of cellular redox: (a) CV obtained from different cell lines. *Source:* Reproduced with permission from Ref. [43]. (b) DPV obtained from different cell lines. *Source:* Reproduced with permission from Ref. [43]. (c) DPV obtained from synchronized cells at different phases of the cell cycle. *Source:* Reproduced with permission from Ref. [10].

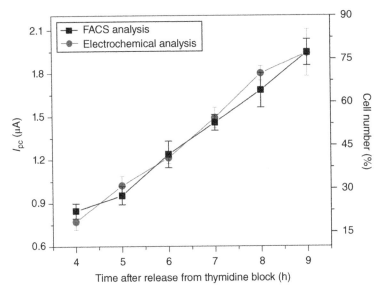

Figure 3.6 Validation of electrochemical measurement of cell viability with standard fluorescent activated cell sorting method. *Source:* Reproduced with permission from Ref. [10].

Electrochemical monitoring of cell proliferation was performed by utilizing the quantification and analysis of voltammetric signals obtained from a cell immobilized platform [2, 3, 6]. Thus, the cell-based platform holds promise for monitoring the healing progress of an active wound. Hence, researchers are focusing on these applications of the cell-based sensor patch. Establishment of cell chip on a biocompatible, bioresorbable, and degradable platform is our current demand for developing a smart wound healing patch for assisting healing as well as monitoring the healing progress [46, 47]. Recently, a chitosan-based sensor patch has been demonstrated for monitoring cell health as well as sensing analytes such as dopamine, serotonin, and glucose in vitro [15, 16]. However, their in vivo application has not been explored yet.

This cell line specific redox peak was subjected to monitoring influences of environmental toxicants and analytes elsewhere [2, 21]. It is well known that living cells have a distinct cell line and cell cycle stage specific redox property [10, 48]. Hence, the environmental influence on cellular redox was employed as a parameter for detection and quantifying of such influences. The changes in peak intensity reflect the state of cell health to any exogenous or endogenous influences affecting the cell viability. Therefore, the presented device is able to monitor environmental toxicants by analyzing and quantifying the redox signal.

The performance of this newly introduced method was verified with the monitoring of cell viability against two potential toxicants such as BPA and DDT where

Table 3.1 Comparison of toxicity analysis between trypan blue exclusion assay and voltammetric peak current analysis.

Toxicant	Concentration (μM)	Viability (%)	I_{pc} (μA)
Bisphenol-A (BPA)	0	100	1.12 ± 0.09
	1	81	0.82 ± 0.05
	2.5	62	0.6 ± 0.09
	5	43	0.31 ± 0.05
	7.5	19	0.15 ± 0.02
Dichlorodiphenyltrichlorehanole (DDT)	0	100	0.81 ± 0.07
	1.5	72	0.56 ± 0.03
	3	59	0.45 ± 0.05
	4.5	46	0.39 ± 0.04
	6	41	0.33 ± 0.02
	7.5	25	0.29 ± 0.02

this fabricated cell chip was applied to detect the toxicity of environmental toxins and showed reciprocal relationship between the signal intensity of redox peaks and the concentrations of BPA and DDT (Table 3.1). This method was also employed for monitoring other environmental toxicants utilizing various cell lines [2, 9, 21]. Later, cell cycle specific redox signals were also employed for monitoring cell cycle stage specific effect of toxicant efficiently [3]. For such case neural cell synchronized at Synthesis and Mitosis phase were exposed to BPA and PCB and electrochemical measurements were performed (Figure 3.7a and b). The concentration dependent reciprocal relationship were observed for both the toxicant indicating that the synchronized cell signals also effective for monitoring toxicity of analytes (Figure 3.7c and d).

This cellular membrane potential is sensitive to applied potentials and thus utilized for controlling several ion gated channels in the cell membrane [49]. This regulated membrane potential was applied for monitoring the endocytosis and exocytosis phenomena. Thus, electrochemically regulated cellular ion channels were applied for many targeted drug delivery and gene therapy applications [50].

3.6 Prospects and Challenges of Cell-Based Sensor

Considering the rapid sensitive detection nature, the developed cell chip-based electrochemical method has been attracted for wider applications in the diagnostic and sensing fields. In particular, the cell-based chip shows very accurate

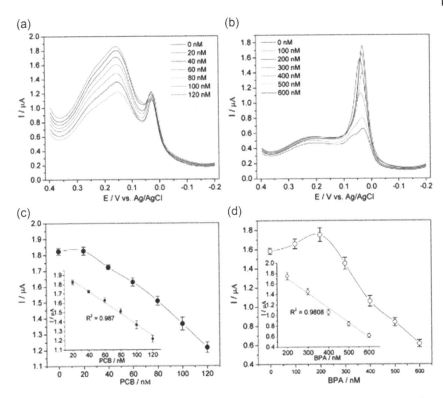

Figure 3.7 Concentration-dependent cytotoxicity: (a) Effect of PCB on cells completely synchronized at G2/M phase and (b) effect of BPA on cells completely synchronized at G1/S phase. Dose–response curve obtained from PCB treatment on G2/M synchronized chip (c) and BPA treatment on G1/S synchronized chip (d).

and reliable responses to newly developed drug or therapeutics [9, 21]. The promising technology also holds promise for real-time monitoring of patient analytes which is critically required for establishing remote monitoring of a patient. Such application requires establishment of a biocompatible platform for cell chip technology for attaining their wearability or implantability [51]. Integration of a remote sensing system on the developed platform is critical for distant monitoring of the patient. The state-of-the-art cell-based devices lack wearability and implantability because they are established on nonconformable rigid carrier substrates (such as Si and ITO glass). In addition, the currently used platforms lack biocompatibility and thus are not suitable for wearable and implantable application. Therefore, the materials with biocompatibility, flexibility, and bioresorbability are our future demand for miniaturization, automation of this promising sensing device.

3.7 Conclusion

Cell-based sensing is an emerging tool for monitoring environmental pollutants or newly developed drug effect studies. This potential tool encompasses a living cell immobilized conductive platform, which is connected to a three-electrode system electrochemical workstation for recording electron exchange phenomenon at a defined potential window. The potential windows were found to be varied from cell line to cell line and even varied between stages of the cell cycle of a same cell line. Later, this potential impulse was found to be enhanced or depressed by any extracellular influence (positive or negative stimuli) based on which effects of various environmental toxicants and newly developed drugs were investigated successfully. Since beginning the cell chip technology faces several challenges, one of which is the establishment of a strong link between the cell and the electrode. Establishment of nanostructured RGD tripeptide molecules overcomes this initial hindrance towards the advancement of this emerging technology. Recently RGD nanostructured modified Au surface was employed to monitor cell cycle stages successfully. Later, these cell line specific electrochemical signals were successfully employed to determine environmental toxicity. These recent advancements of the cell chip technology have been exploring their future application to the real-time monitoring of clinical analytes. But these versatile applications are still facing challenges due to the lack of flexibility of the commonly used cell chip platform which is an essential requirement for their application to a living body or robotics. In addition, the integration of a microfluidic system with the flexible cell support for the sample acquisition and processing prior exposing cell on chip will further enhance the possibility of the real-time monitoring clinical analytes. As future work, the cell chip technology has a promising perspective to be combined with other sensing materials such as graphene [52] and other lab-on-chip technologies [53–55].

References

1 Bery, M.N. and Grivell, M.B. (1995). *Bio-electrochemistry of Cells and Tissues* (eds. D. Walz, H. Berry and G. Milazzo), 134–158. Birkhauser, Basel: Verlag.

2 Kafi, M.A., Kim, T.H., Yagati, A.K. et al. (2010). Nanoscale fabrication of a peptide layer in cell chip to detect effects of environmental toxins on HEK293 cells. *Biotechnology Letters* 32: 1797–1802.

3 Kafi, M.A., Yea, C.H., Kim, T.H. et al. (2013). Electrochemical cell chip to detect environmental toxicants based on cell cycle arrest technique. *Biosensors and Bioelectronics* 41: 192–198.

4 N'Dri, N.A., Shyy, W., and Tran-Son-Tay, R. (2003). Computational modeling of cell adhesion and movement using a continuum-kinetics approach. *Biophysical Journal* 85: 2273–2286.

5 Discher, D.E., Mooney, D.J., and Zandstra, P.W. (2009). Growth factors, matrices, and forces combine and control stem cells. *Science* 324: 1673–1677.

6 Kafi, M.A., Kim, T.H., Yea, C.H. et al. (2010). Effects of nanopatterned RGD peptide layer on electrochemical detection of neural cell chip. *Biosensors and Bioelectronics* 26: 1359–1365.

7 Kim, T.H., El-Said, W.A., An, J.H., and Choi, J.W. (2013). ITO/gold nanoparticle/RGD peptide composites to enhance electrochemical signals and proliferation of human neural stem cells. *Nanomedicine* 9: 336–344.

8 Angnes, L., Richter, E.M., Augelli, M.A., and Kume, G.H. (2000). Gold electrodes from recordable CDs. *Analytical Chemistry* 72 (21): 5503–5506.

9 El-Said, W.A., Yea, C.H., Kim, H. et al. (2009). Cell-based chip for the detection of anticancer effect on HeLa cells using cyclic voltammetry. *Biosensors and Bioelectronics* 24: 1259–1265.

10 Kafi, M.A., Kim, T.H., An, J.H., and Choi, J.W. (2011). Fabrication of cell chip for detection of cell cycle progression based on electrochemical method. *Analytical Chemistry* 83: 2104–2111.

11 Kafi, M.A., El-Said, W.A., Kim, T.H., and Choi, J.W. (2012). Cell adhesion, spreading, and proliferation on surface functionalized with RGD nanopillar arrays. *Biomaterials* 33: 731–739.

12 Lupu, S., Lete, C., Marin, M. et al. (2009). Electrochemical sensors based on platinum electrodes modified with hybrid inorganic–organic coatings for determination of 4-nitrophenol and dopamine panel. *Electrochimica Acta* 54: 1932–1938.

13 Lee, J.Y., Connor, S.T., Cui, Y., and Peumans, P. (2008). Solution-processed metal nanowire mesh transparent electrodes. *Nano Letters* 8: 689–692.

14 Yoo, S.H. and Park, S. (2007). Platinum-coated, nanoporous gold nanorod arrays: synthesis and characterization. *Advanced Materials* 19: 1612–1615.

15 Kafi, M.A., Paul, A., Vilouras, A., and Dahiya, R. (2018). Chitosan-graphene oxide based ultra-thin conformable sensing patch for cell-health monitoring. In: *IEEE Sensors*, 1–4. IEEE.

16 Vilouras, A., Paul, A., Kafi, M.A., and Dahiya, R. (2018). Graphene oxide-chitosan based ultra-flexible electrochemical sensor for detection of serotonin. In: *IEEE sensors*, 1–4. IEEE.

17 Wang, G., Zhang, L., and Zhang, J. (2012). A review of electrode materials for electrochemical super capacitors. *Journal of Chemical Society Review* 41: 797–828.

18 Felipe, A., Cruz, D., Norena, N. et al. (2014). A low-cost miniaturized potentiostat for point-of-care diagnosis. *Biosensors and Bioelectronics* 62: 249–254.

19 Yin, M.J., Huang, B., Gao, S. et al. (2016). Optical fiber LPG biosensor integrated microfluidic chip for ultrasensitive glucose detection. *Biomedical Optics Express* 7: 2067–2077.

20 Libis, V., Delépine, B., and Faulon, J.L. (2016). Expanding biosensing abilities through computer-aided design of metabolic pathways. *American Chemical Society Synthetic Biology* 5: 1076–1085.

21 Kafi, M.A., Kim, T.H., An, J.H., and Choi, J.W. (2011). Electrochemical cell-based chip for the detection of toxic effects of bisphenol-A on neuroblastoma cells. *Biosensors and Bioelectronics* 26: 3371–3375.

22 Wang, S. and Du, D. (2004). Differential pulse voltammetry determination of ascorbic acid with ferrocene-l-cysteine self-assembled supramolecular film modified electrode. *Sensors and Actuators B: Chemical* 97: 373–378.

23 Chaki, N.K. and Vijayamohanan, K. (2002). Self-assembled monolayers as a tunable platform for biosensor applications. *Biosensors and Bioelectronics* 17: 1–12.

24 Yao, Y., Yu, J., and Chen, X. (2014). Study on the optically transparent near-field and far-field RFID reader antenna. *International Journal of Antennas and Propagation*: 149051, 5 pages. http://dx.doi.org/10.1155/2014/149051.

25 Shen, S.P. and Dow, W.P. (2014). Adhesion enhancement of a plated copper layer on an AlN substrate using a chemical grafting process at room temperature. *Journal of the Electrochemical Society* 161: D579–D585.

26 Li, Y., Johnson, R.W., Zhang, R. et al. (2010). Ti/Au die backside metallization for flip chip heat spreader attachment. *IEEE Transactions on Electronics Packaging Manufacturing* 33: 44–54.

27 Yang, M., Yi, X., Wang, J., and Zhou, F. (2014). Electroanalytical and surface plasmon resonance sensors for detection of breast cancer and Alzheimer's disease biomarkers in cells and body fluids. *The Analyst* 139: 1814–1825.

28 Liu, J.S., Xiao, Q.L., Ge, D. et al. (2015). A microfluidic chip with integrated microelectrodes for real-time dopamine detection. *Chinese Journal of Analytical Chemistry* 43: 977–982.

29 Cheah, B.C., Macdonald, A.I., Martin, C. et al. (2016). An integrated circuit for chip-based analysis of enzyme kinetics and metabolite quantification. *IEEE Transactions on Biomedical Circuits and Systems* 10 (3): 721–730. https://doi.org/10.1109/TBCAS.2015.2487603.

30 Lei, K.M., Heidari, H., Mak, P.I. et al. (2017). A handheld high-sensitivity micro-NMR CMOS platform with B-field stabilization for multi-type biological/chemical assays. *IEEE Journal of Solid-State Circuits* 52: 284–297.

31 Jackson, N., Sridharan, A., Anand, S. et al. (2010). Long-term neural recordings using MEMS based movable microelectrodes in the brain. *Frontiers in Neuroengineering* 3: 10. https://doi.org/10.3389/fneng.2010.00010l.

32 Huang, Y., Williams, J.C., and Johnson, S.M. (2012). Brain slice on a chip: opportunities and challenges of applying microfluidic technology to intact tissues. *Lab on a Chip* 12: 2103–2117.

33 Bucher, V., Brunner, B., Leibrock, C. et al. (2001). Electrical properties of a light-addressable microelectrode chip with high electrode density for extracellular stimulation and recording of excitable cells. *Biosensors and Bioelectronics* 16: 205–210.

34 Pecqueur, S., Vuillaume, D., and Alibart, F. (2018). Perspective: organic electronic materials and devices for neuromorphic engineering. *Journal of Applied Physics* 124: 151902.

35 Soussou, W.V., Yoon, G.J., Brinton, R.W., and Berger, T.W. (2007). Neuronal network morphology and electrophysiology of hippocampal neurons cultured on surface-treated multielectrode arrays. *IEEE Transactions on Biomedical Engineering* 54: 1309–1320.

36 Dwyer, D.S., Liu, Y., and Bradley, R.J. (1999). An ethanol-sensitive variant of the PC12 neuronal cell line: sensitivity to alcohol is associated with increased cell adhesion and decreased glucose accumulation. *Journal of Cellular Physiology* 178: 93–101.

37 Kleinman, H., Klebe, R., and Martin, G. (1981). Role of collagenous matrices in the adhesion and growth of cells. *Journal of Cell Biology* 88: 473–485.

38 Yea, C.H., Lee, B., Kim, H. et al. (2008). The immobilization of animal cells using the cysteine-modified RGD oligopeptide. *Ultramicroscopy* 108: 1144–1147.

39 Yea, C.H., Min, J., and Choi, J.W. (2007). The fabrication of cell chips for use as bio-sensors. *BioChip Journal* 1: 219–227.

40 Ruoslahti, E. (1996). RGD and other recognition sequences for integrins. *Annual Review of Cell and Developmental Biology* 12: 697–715.

41 El-said, W.A., Kim, T.H., Kim, H., and Choi, J.W. (2010). Detection of effect of chemotherapeutic agents to cancer cells on gold nanoflower patterned substrate using surface-enhanced Raman scattering and cyclic voltammetry. *Biosensors and Bioelectronics* 26: 1486–1492.

42 Hersel, U., Dahmen, C., and Kessler, H. (2003). RGD modified polymers: biomaterials for stimulated cell adhesion and beyond. *Biomaterials* 24: 4385–4415.

43 Huang, J., Grater, S.V., Corbellini, F. et al. (2009). Impact of order and disorder in RGD nanopatterns on cell adhesion. *Nano Letters* 9: 1111–1116.

44 Kafi, M.A., Kim, T.H., Lee, T., and Choi, J.W. (2013). Cell chip with nano-scale peptide layer to detect dopamine secretion from neuronal cells. *Journal of Nanoscience and Nanotechnology* 11: 7086–7090.

45 Liu, Q., Wu, C., Cai, H. et al. (2014). Cell-based biosensors and their application in biomedicine. *Chemical Reviews* 114: 6423–6461.

46 Brown, M.S., Ashley, B., and Koh, A. (2018). Wearable technology for chronic wound monitoring: current dressings, advancements, and future prospects. *Frontiers in Bioengineering and Biotechnology* 6: 47.

47 Sousa, M.P., Neto, A.I., Correia, T.R. et al. (2018). Bioinspired multilayer membranes as potential adhesive patches for skin wound healing. *Biomaterials Science* 6: 1962–1975.

48 Kafi, M.A., Kim, T.H., and Choi, J.W. (2011). Cell chip to analyze cell lines and cell cycle stages based on electrochemical method. In: *SENSORDEVICES 2011: The Second International Conference on Sensor Device Technologies and Applications*, 147–150. IARIA.

49 Alberts B, Johnson A, Lewis J, et al. (2002). *Molecular Biology of the Cell*. 4th edition. New York: Garland Science. Ion Channels and the Electrical Properties of Membranes. Available from:https://www.ncbi.nlm.nih.gov/books/NBK26910/

50 Li, T. and Chen, J. (2018). Voltage-gated sodium channels in drug discovery. In: *Ion Channels in Health and Sickness* (ed. F.S. Kaneez). IntechOpen http://dx.doi.org/10.5772/intechopen.78256.

51 Berg, A.V., Mummery, C.L., Passiera, R., and Meer, A.D.V. (2019). Personalized organs-on-chips: functional testing for precision medicine. *Lab on a Chip* 19: 198.

52 Parvizi, R., Azad, S., Dashtian, K. et al. (2019). Natural source-based graphene as sensitising agents for air quality monitoring. *Scientific Reports* 9: 3798.

53 Shah S and Heidari H, "On-chip magnetoresistive sensors for detection and localization of paramagnetic particles," In *2017 IEEE SENSORS, Glasgow*, 2017, pp. 1–3. IEEE. doi: 10.1109/ICSENS.2017.8233894.

54 Yin, Z., Bonizzoni, E., and Heidari, H. (2018). Magnetoresistive biosensors for on-chip detection and localization of paramagnetic particles. *IEEE Journal of Electromagnetics, RF and Microwaves in Medicine and Biology* 2 (3): 179–185. https://doi.org/10.1109/JERM.2018.2858562.

55 Nabaei, V., Chandrawati, R., and Heidari, H. (2018). Magnetic biosensors: modelling and simulation. *Biosensors and Bioelectronics* 103: 69–86.

4

Electronic Tongues

Flavio M. Shimizu[1], Maria Luisa Braunger[2], Antonio Riul, Jr.[2], and Osvaldo N. Oliveira, Jr.[3]

[1] *Brazilian Nanotechnology National Laboratory (LNNano), Brazilian Center for Research in Energy and Materials (CNPEM), Campinas, São Paulo, Brazil*
[2] *Department of Applied Physics, "Gleb Wataghin" Institute of Physics, University of Campinas (UNICAMP), Campinas, São Paulo, Brazil*
[3] *São Carlos Institute of Physics, University of São Paulo (USP), São Carlos, São Paulo, Brazil*

4.1 Introduction

An e-tongue is a multisensory array based on electroanalytical methods and statistical techniques that have been applied in qualitative and/or quantitative discrimination of beverages (coffee [1, 2], wine [3–5], tea [6], water [7, 8]), foodstuff [9–11], pesticides [12–15], and suppression effects [16–18]. The main reason to resort to ETs arises in situations where the use of a human panel is not possible, such as in the continuous monitoring of industrial processes, analysis of unpleasant or poisonous/hazardous samples (drugs, virus, bacteria, pollutants), and for economical limitations [19]. The basic idea in an ET is to mimic the biological system, as illustrated in Figure 4.1. Taste buds in the human tongue (1) when in contact with foodstuffs (2) send a flood of electrical stimulus to the brain (3), which has the unique capability to group them into specific patterns to classify tastes (4). Likewise, an array of sensing units made with electrodes modified with materials having diverse chemical compositions (5) immersed in liquids (6) create a fingerprint of the sample through different electroanalytical responses. Information can be processed using statistical and computational techniques (7) to identify a sample (or a taste) (8).

Pioneering works from Toko [20], Legin, Vlasov, and coworkers [21–23] back in the 1990s established the use of ETs exploiting nonspecific interactions. A few

Smart Sensors for Environmental and Medical Applications, First Edition. Edited by Hamida Hallil and Hadi Heidari.
© 2020 The Institute of Electrical and Electronics Engineers, Inc.
Published 2020 by John Wiley & Sons, Inc.

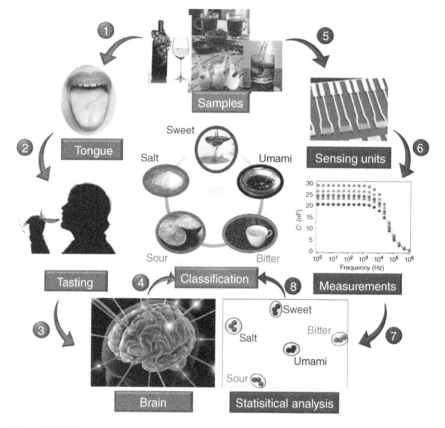

Figure 4.1 Schematic chart of the e-tongue working principle.

years later the use of biomolecules [24] (enzymes, proteins, cells) as sensing materials represented a breakthrough in extending the e-tongue concept, sometimes referred as bioelectronic tongue (bET). With specific interactions new achievements could be reached in taste assessment [25, 26]. Despite the good similarity achieved with human sensory panels, effectively mimicking the human taste transduction, key points like drift in sensor signals caused by aging and/or surface contamination, simpler, more sensitive systems, reproducibility, accuracy, and reliability need to be improved. Reviews exist on a variety of topics related to e-tongues. They discuss the real capability of e-tongues and their use in biosensing [19], electrochemical sensors using distinct types of working electrodes, electrode cleaning methods, porphyrin-based films [27], feature data selection, preprocessing and data compression, and pattern recognition methods for classification and prediction [28]. Overall, contributions about e-tongues can be classified roughly into: (i) novel materials and nanoarchitectonics [29] for sensing layers; (ii) statistical methods for data mining and machine learning; (iii) new design of electrodes or measurement systems.

In this chapter, we describe recent advances with special emphasis on microfluidic platforms based on polymers and paper, new designs, and new sensing and biosensing applications.

4.2 General Applications of E-tongues

The systems to be discussed here are based on electrochemical sensors (potentiometry, electrochemical impedance spectroscopy, and voltammetry), field-effect transistors (FETs), and impedimetric measurements. Overall, potentiometry [30] is the most applied characterization in e-tongue systems, with voltammetric methods being normally less influenced by electrical disturbances than are potentiometric and impedimetric measurements. However, they are still limited to redox-active substances. In contrast, some FETs and impedimetric devices do not require conventional electrochemical cells and electrochemical activity. Therefore, we emphasize that each system has its own particularity with advantages and disadvantages.

Manzoli et al. [31] tailored polyaniline (PANi) properties for sensing layers of ETs by synthesizing AgCl-PANI nanocomposites with no polymer de-doping at high pHs. Correa et al. [32] decorated polymeric nanofibers with 1D and 2D materials to tune the sensor properties for food and agricultural applications of ETs. Electrodes were modified with metallic films (Ir, Rh, Ag, Pd, Pt, and Au) to monitor the process of yogurt culture and storage [33], antioxidant capacity in *aliso* dilutions [34], honey [35, 36], mineral content in spring water [8], where the voltammetric data were analyzed with principal component analysis (PCA), partial least square (PLS), and linear discriminant analysis (LDA). Smart nanomaterials [37, 38] have been applied as bET for determining hydrogen peroxide, ethanol, and carbohydrates (glucose, xylose, galactose, mannose), with low limits of detection (LoD) (down to $10^{-8}\,mol\,l^{-1}$) and responses comparable to conventional high-performance liquid chromatography (HPLC) systems.

Wine is one of the beverages most studied using e-tongues. Metal oxide nanoparticles (CeO_2, NiO, and TiO_2) have been used in sensing layers to monitor chemical changes during ripening of the grapes [39]. A compact, low-power-consumption portable e-tongue composed by six ion-selective FETs sensitive to pH, Na^+, K^+, Ca^{2+}, Cl^-, and $[CO]_3{}^{2-}$, was applied in the analysis of 78 Cava wine samples to classify the samples according to the aging time (96% prediction of correction using PLS), and quantify the total acidity, pH, volumetric alcoholic degree, potassium, conductivity, glycerol, and methanol parameters in a period of 150 hours [40]. Still regarding wine analysis, a mini-review [41] and a critical [42] review elaborate on electrochemistry monitoring with e-tongue systems.

The use of electroactive species to provide high specificity in pharmaceuticals enabled the e-tongue application to detect atenolol and propranolol (antiarrhythmic medicines) [43], insulin [44], diclofenac (in the presence of paracetamol and naproxen) [45], and amino acids (tryptophan, tyrosine and cysteine) [46]. A hybrid e-tongue [47] (hET) using a flow injection system equipped with two electrochemical sensors and one optical sensor could evaluate bitterness in soft drinks fortified with plant extracts of green tea. Zabadaj et al. [48] combined potentiometric and voltammetric detection to investigate yeast culture media during batch fermentation, obtaining good correlation when data treated with PLS were compared with HPLC results. Commercial e-tongue, e-nose, near infrared spectrum (NIR), and mid-infrared spectrum (MIR) analysis were applied in raw honey samples (Vitex, Jujube, and Acacia). Partial least squares discriminant analysis (PLS-DA), support vector machine (SVM) algorithms, and Interval partial least squares (iPLS) model were employed to classify the botanical origin of honeys. The e-tongue system was more sensitive to amino acids, minerals, phenols, monosaccharides, and disaccharides existing in honey [49], presenting simpler data processing and best performance for adulterated honey using PCA. The first totally integrated, easy-to-use, and portable electrochemical platform integrating a smartphone and homemade potentiostat for point-of-use assays successfully distinguished Brazilian honey samples according to their botanical and geographic origins. The system presented great simplicity, high-electrochemical performance detection (linear sweep, cyclic, and square wave voltammetry), 6 hours autonomy, low-cost, portability, and wireless communication. Moreover, an App for Android operational system (named TongueMetrix) was developed to collect, store, and process multivariate data [50]. Cavanillas et al. [51] proposed a self-polishing voltammetric sensor to urea, milk, and sewage water sample analysis, achieving high correlation due to the regeneration of the electrode surface, avoiding the accumulation of redox reaction products at the electroactive surface during the data acquisition.

As already mentioned, chemometric tools play an important role in e-tongue pattern recognition because they are responsible for translating hundreds or thousands of multivariate signals onto a visualization map [52]. This can be done qualitatively or quantitatively by means of PLS [53, 54] or artificial neural network (ANN) [4, 55]. Indeed, methods to reduce high-dimensional data down to two or three dimensions are ubiquitous in statistical data analysis, minimizing loss of information with techniques such as PCA [56], Fastmap [52], wavelet- [57], and fast Fourier-transform (FFT) [58], which offer better results as shown in next examples.

Legin et al. [59] reported a case study of quantitative analysis of microcystin concentration in water samples, with the samples being classified as either toxic or nontoxic using a potentiometric e-tongue. This task was only possible with multi-variate standardization of mathematical correction to solve drift problems over the period of algae growth and proliferation that demand recalibration process of a sensor array. Procedures such as these are usually not feasible due to the complexity and investments required. Pérez-Ràfols et al. [60] employed PLS to resolve overlapped signals, with satisfactory results in the simultaneous determination of Cd^{2+}, Pb^{2+}, Tl^{1+}, and Bi^{3+} in the presence of Zn^{2+} and In^{3+} in a spiked tap water sample. del Valle et al. reported several works of voltammetric e-tongues combined with ANNs to quantify individual species [4, 5] and ternary mixtures [57]. Cetó et al. [61] employed FFT for the compression and reduction of signal dimensionality, and the coefficients were applied as inputs to build qualitative and quantitative models using either LDA or PLS regression to discriminate the ageing time of wine in barrels and predict global scores assigned by a sensory panel. Although not replacing experts, it proved to be a valuable tool to overcome the lack of knowledge of compounds responsible for some wine sensations.

4.3 Bioelectronic Tongues (bETs)

The use of biomolecules in e-tongues began in the late 1990s [24]; the "bio electronic tongue" term was coined in 2005 by Tonning et al. [62]. Bio e-tongues have been used in water analysis; to discriminate varieties of grapes [63] through phenolic and sugar content; to detect phytic acid [64, 65]; for the diagnosis of Chagas and Leishmaniasis diseases [66]; and to detect catechol [12], glucose [67], and triglycerides [68]. Potentiometric e-tongues were utilized to determine urea and creatinine in urine [69] and urinary malfunctions and creatinine levels [70], and in the detection of prostate cancer [71]. For the latter work, the e-tongue consisted of an array of seven working electrodes (Ir, Rh, Pt, Au, Ag, Co, and Cu) housed in a stainless-steel cylinder employed as the counter electrode. With PLS analysis they achieved a 91% sensitivity and 73% specificity to distinguish the urine of cancer patients from the urine in healthy patients. More recently, a voltammetric e-tongue made with glassy carbon (GC), gold (Au), platinum (Pt), silver (Ag), nickel (Ni), palladium (Pd), and copper (Cu) electrodes was applied to identify the urinary creatinine content (UCC) from 59 volunteers without dilution. PCA and SVMs were employed as classification methods, identifying the urine samples into three classes according to low, medium, and high creatinine levels [72].

A molecular imprinted polymer (MIP)-based e-tongue was applied in classifying five proteins (cytochrome c, ribonuclease A, α-lactalbumin, albumin, and myoglobin). Despite the binding capability of MIP to all four proteins, there were no differences in PCA patterns for the MIP and nonimprinted polymer (NIP)-based chemosensor arrays [73]. An e-tongue approach based on MIPs integrated in a multielectrode array was used with differential pulse voltammetry (DPV) for detecting 4-ethylphenol (4-EP) and 4-ethylguaiacol (4-EG) compounds normally present in wines. Graphite was used as the conducting material and MIP particles were incorporated in the electrode using the sol-gel technique to form MIP-functionalized electrodes. PCA was used for identification and ANN for quantitative analysis, obtaining a correlation coefficient >0.98, though authors reported a lack of sensitivity of MIPs [74].

Although natural characteristics of biological systems cannot be imitated, such as the substitution of old taste buds when specific recognition is required [75], an exquisite bio(e-tongue) was developed by immobilizing taste buds on microelectrode arrays, with the principle of detection based on receptor cell stimulation activated by NaCl [76] and acidic [77] solutions. In this trend, Park et al. [75] employed nanovesicles containing human taste receptors, proteins heterodimeric G-protein-coupled receptors (GPCRs) composed of human taste receptor type 1 member 2 (hTAS1R2), and human taste receptor type 1 member 3 (hTAS1R3) for the detection of sucrose, fructose, aspartame, and saccharin. Despite the high similarity with the human tongue in selectivity, the bET did not respond to tasteless sugars (cellobiose and D-glucuronic acid) as shown in Figure 4.2. Later on, Park's group published [78] a duplex bio(e-tongue) to sense umami and sweet tastes simultaneously using the same method. Lee et al. [79] demonstrated that an insect receptor (10 of Apis mellifera [AmGr10]) in nanovesicles may be used to determine umami substances. The reasons to use nanovesicles stem from advantages of protecting the proteins in the sensing units.

4.4 New Design of Electrodes or Measurement Systems

Microfluidic systems appeared after Daikuzono et al. [80] proved the successful usage of the first microfluidic e-tongue design (Figure 4.3a). They reported on a microfluidic e-tongue formed by gold interdigitated electrodes (IDEs) inside poly(dimethylsiloxane) (PDMS) microchannels. The IDEs were modified with organic films deposited by the LbL technique (Figure 4.3b), thus forming the sensing units of the device. Impedance measurements were taken as it is a noninvasive method and avoids polarization effects in the samples, allowing for faster measurements. Postprocessing of raw data by statistical methods provides a visual representation and ease interpretation of the results. As an example, $1 \, \text{mmol} \, l^{-1}$ aqueous solutions of HCl (sour), NaCl (salty), caffeine (bitter), sucrose (sweet), and L-glutamic acid monosodium salt hydrate (umami) were analyzed with good distinction of all tastes at 1 kHz. Latter it was applied to evaluate the quality of coffee [82], and presence of gliadin [11] (a gluten protein) of gluten-free and gluten-containing foodstuff samples. The system was able to discriminate gliadin

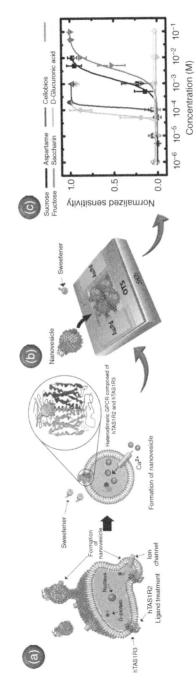

Figure 4.2 Schematic diagram of (a) nanovesicle containing GPCRs fabrication, (b) surface modification of electrode surface, and (c) response of bET to various sweeteners (artificial, natural, and tasteless). *Source*: Reproduced with permission from [75]. Copyright 2015, American Chemical Society.

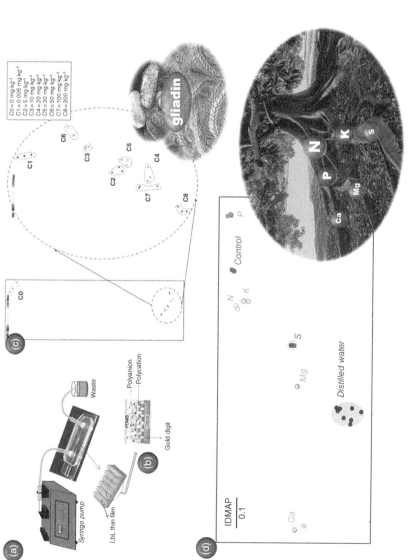

Figure 4.3 (a) Schematic view of the experimental setup used in the dynamic LbL assembly. (b) Cross-section view of the LbL film inside the microchannel. (c) IDMAP plot for distinguishing gliadin solutions at various concentrations. *Source:* Adapted and reproduced with permission from [11]. Copyright 2017, American Chemical Society. (d) IDMAP plot of capacitance data with selected frequencies for detection of soil samples enriched with N, P, K, Ca, Mg, and S. The black scale bar at the bottom of graph is equivalent to 0.1 in Euclidean metrics and can be used to measure the distance between the clusters formed. *Source:* Reproduced with permission from [81]. Copyright 2017, MDPI.

samples down to 0.005 mg kg^{-1} (trace amounts) in contaminated gluten-free food-stuff. The high performance was achieved by means of a statistical feature selection using multidimensional projection analysis (Interactive Document Map [IDMAP]) to process the raw data, as shown in Figure 4.3c. Further experiments described by Braunger et al. depict the detection of soil samples enriched with plant macronutrients (N, P, K, Ca, Mg, and S) in a proof-of-concept experiment to support precision agriculture [81]. Among the statistical tools used, an optimized performance was obtained with IDMAP for selected frequency ranges of the raw data (Figure 4.3d). Distinction of all soil samples was also possible with the simple and linear PCA, for measurements taken at a fixed single frequency. Finally, authors indicated good reproducibility and reuse of the e-tongue system after a simple water cleansing, with no signs of cross-contamination from the soil samples.

A novel paper-based e-tongue integrated with Ag/AgCl reference and using only 40 µl for sampling paved the way for miniaturization and low fabrication cost. The paper-based e-tongue was able to distinguish 34 beers from brands, and types, indicating also the presence of stabilizers and antioxidants, dyes, or even unmalted cereals and carbohydrates added to the fermentation wort (Figure 4.4). The system could also classify beers according to the type of fermentation (low, high), predict the pH and in part the alcohol content of the samples, in addition to differentiate wine samples produced from different varieties of grapes [83]. Reports have also appeared of a low-cost paper-based ET comprising four paper-based potentiometric sensors and an Ag/AgCl reference electrode. The sensitivity is not high, but it is sufficient to verify adulteration of 14 different water samples from natural springs in Brazil, even in cases of possible adulteration [84].

Gaál et al. explored the 3D-printing fused deposition modeling (FDM) (Figure 4.5a) as an alternative to PDMS to fabricate microchannels using polylactic acid (PLA) (Figure 4.5b) [85]. LbL films were deposited onto IDEs placed inside the 3D-printed microchannels, with the system tested in a proof-of-concept experiment similar to that performed by Daikuzono et al. [80]. The 3D printed system was able to discriminate sucrose, NaCl, HCl, and caffeine solutions below the human threshold (1 mmol l^{-1}), integrating a cost-effective technology that allows for alternative materials for fast prototyping of complex structures in microfluidic systems. In another study, Gaál et al. [86] presented a 3D-printed e-tongue applied to soil analysis, also using LbL films deposited onto IDEs as sensing units, using the same soil samples presented by Braunger et al. [81]. Instead of gold IDEs, the IDEs were 3D printed using graphene-based PLA filaments (Figure 4.5c), in a home-made FMD 3D-printer. The PCA plot indicated good distinction of all tested soil samples.

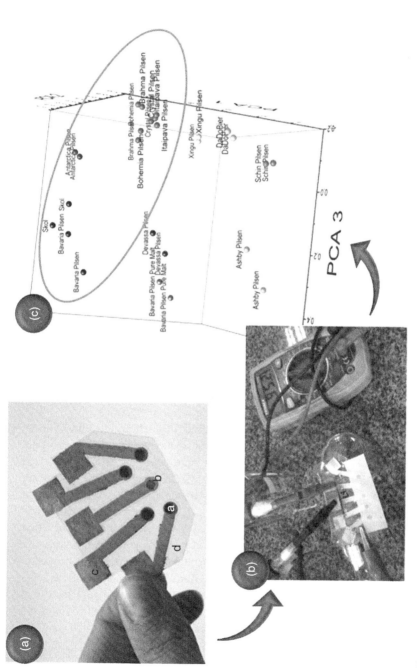

Figure 4.4 PCA scores plot for Pilsner beer samples. (a) Photograph of paper-based electronic tongue with an integrated reference: a is the working electrodes, b the Ag/AgCl reference electrode, c the electrical contacts and d the lamination foiL. (b) Schematic picture of electronic tongue detection system performed by means of a multimeter. (c) PCA scores plot for Pilsner beer samples. PC 1, 2 and 3 described 41, 36 and 14% of total information, respectively. *Source:* Reproduced with permission from [83]. Copyright 2016, Elsevier.

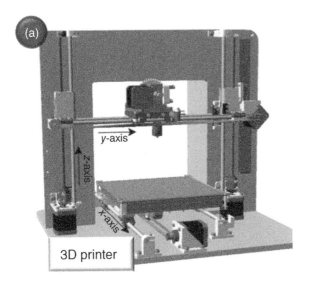

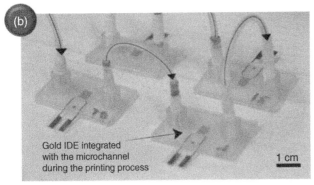

Figure 4.5 (a) Schematic chart of fused deposition modeling 3D printer. (b) 3D printed microchannel with gold IDEs interconnected to form a microfluidic e-tongue. *Source:* Reproduced with permission from [85]. Copyright 2017, Elsevier. (c) Fully 3D printed IDE with graphene-based PLA filaments. *Source:* Reproduced with permission from [86]. Copyright 2018, Gaál et al.

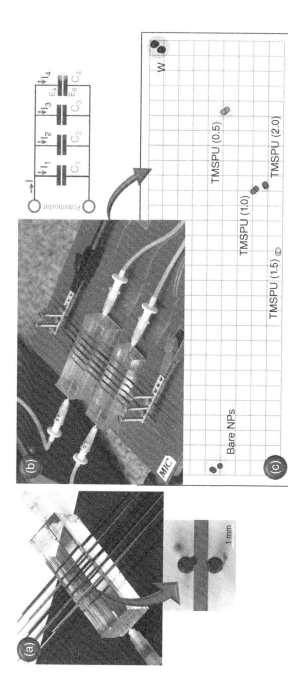

Figure 4.6 (a) Picture of microfluidic e-tongue based on a single chip of PDMS prepared with PSR method [89], with electrodes coated by different metal and oxide films, and a zoomed image showing the cross-section of microfluidic channel where electrodes are inserted above and below it (red dye solution was used for sake of visualization). (b) Multiplex measurement system with connectors short-circuiting the electrodes. (c) IDMAP plot of SiO$_2$ nanoparticles with different functionalization level of TMSPU. *Source:* Reproduced with permission from [88]. Copyright 2018, American Chemical Society.

A key limitation on e-tongue systems is the replacement of sensing units during analysis, but that may be overcome with the work by Shimizu et al. [87, 88]. They developed a new microfluidic e-tongue consisting of a single PDMS piece (avoiding photolithographic process) having stainless steel microtubes coated with different metals and oxides (gold, platinum, nickel, iron, and aluminum) as sensing units, illustrated in Figure 4.6a. Striking features are the use of renewable sensing units, high reproducibility by simply displacing the wires in the single-response analysis (Figure 4.6b) by short-circuiting the electrodes. This e-tongue system can be used to monitor the properties of nanomaterials, e.g. with distinct functionalization or oxidation levels. Figure 4.6c shows the IDMAP plot discriminating silica nanoparticles (SiO_2) functionalized with different levels of 1-[3-(trimethoxysilyl)propyl]urea (TMSPU), confirming the efficacy in the use of an e-tongue.

4.5 Challenges and Outlook

E-tongues have evolved into a powerful tool within a short period (~30 years). However, albeit commercial systems are available (Alpha MOS and Insent), the technology so far has not reached the market in a substantial manner. We hope that this chapter motivates readers to develop collaborative efforts from different areas to overcome the barriers that prevent commercialization. The challenges that are being addressed include: (i) miniaturization and integration with emergent technologies aiming at optimized systems for commercial applications; (ii) fabrication of smart sensors [90] and use of machine learning processes are a reality not only to deal with the Big Data paradigm, but also to find out solutions and integration with the Internet of Things (IOT). The inclusion of biomolecules brought a novel concept on clinical diagnosis. Oliveira et al. [91–94] have directed efforts in the detection of different cancer biomarkers (breast, pancreas, head–neck) using different surface modifications with nanostructured platforms to increase the sensitivity of label-free immunosensors by means of electrical and electrochemical measurements, assisting the combat against cancer for a better living for human beings.

Acknowledgments

The authors are grateful for the financial support of the Brazilian funding agencies: FAPESP (Grant No. 2013/14262-7, 2015/14836-9, 2017/11277-4), CAPES, and CNPq.

References

1 Riul, A., Dantas, C.A.R., Miyazaki, C.M., and Oliveira, O.N. Jr. (2010). Recent advances in electronic tongues. *Analyst* 135: 2481–2495.

2 Flambeau, K.J., Lee, W.-J., and Yoon, J. (2017). Discrimination and geographical origin prediction of washed specialty Bourbon coffee from different coffee growing areas in Rwanda by using electronic nose and electronic tongue. *Food Science and Biotechnology* 26: 1245–1254.

3 Riul, A. Jr., de Sousa, H.C., Malmegrim, R.R. et al. (2004). Wine classification by taste sensors made from ultra-thin films and using neural networks. *Sensors and Actuators B: Chemical* 98: 77–82.

4 González-Calabuig, A. and del Valle, M. (2018). Voltammetric electronic tongue to identify Brett character in wines. On-site quantification of its ethylphenol metabolites. *Talanta* 179: 70–74.

5 Ceto, X., Capdevila, J., Minguez, S., and del Valle, M. (2014). Voltammetric bioelectronic tongue for the analysis of phenolic compounds in rosé cava wines. *Food Research International* 55: 455–461.

6 Ghosh, A., Tamuly, P., Bhattacharyya, N. et al. (2012). Estimation of theaflavin content in black tea using electronic tongue. *Journal of Food Engineering* 110: 71–79.

7 Oliveira, J.E., Grassi, V., Scagion, V.P. et al. (2013). Sensor array for water analysis based on interdigitated electrodes modified with fiber films of poly(lactic acid)/multiwalled carbon nanotubes. *IEEE Sensors Journal* 13: 759–766.

8 Carbó, N., Carrero, J.L., Garcia-Castillo, F.J. et al. (2018). Quantitative determination of spring water quality parameters via electronic tongue. *Sensors* 18: 40.

9 Apetrei, I.M. and Apetrei, C. (2016). Application of voltammetric e-tongue for the detection of ammonia and putrescine in beef products. *Sensors and Actuators B: Chemical* 234: 371–379.

10 Chaibun, T., La-o-vorakiat, C., O'Mullane, A.P. et al. (2018). Fingerprinting green curry: an electrochemical approach to food quality control. *ACS Sensors* 3: 1149–1155.

11 Daikuzono, C.M., Shimizu, F.M., Manzoli, A. et al. (2017). Information visualization and feature selection methods applied to detect gliadin in gluten-containing foodstuff with a microfluidic electronic tongue. *ACS Applied Materials & Interfaces* 9: 19646–19652.

12 Aoki, P.H.B., Alessio, P., Furini, L.N. et al. (2013). Molecularly designed layer-by-layer (lbl) films to detect catechol using information visualization methods. *Langmuir* 29: 7542–7550.

13 Scagion, V.P., Mercante, L.A., Sakamoto, K.Y. et al. (2016). An electronic tongue based on conducting electrospun nanofibers for detecting tetracycline in milk samples. *RSC Advances* 6: 103740–103746.

14 Facure, M.H.M., Mercante, L.A., Mattoso, L.H.C., and Correa, D.S. (2017). Detection of trace levels of organophosphate pesticides using an electronic tongue based on graphene hybrid nanocomposites. *Talanta* 167: 59–66.

15 Oliveira, J.E., Scagion, V.P., Grassi, V. et al. (2012). Modification of electrospun nylon nanofibers using layer-by-layer films for application in flow injection electronic tongue: detection of paraoxon pesticide in corn crop. *Sensors and Actuators B: Chemical* 171–172: 249–255.

16 Dittrich, P.S. and Manz, A. (2006). Lab-on-a-chip: microfluidics in drug discovery. *Nature Reviews. Drug Discovery* 5: 210–218.

17 Hoffmeister, C.R.D., Fandaruff, C., da Costa, M.A. et al. (2017). Efavirenz dissolution enhancement III: colloid milling, pharmacokinetics and electronic tongue evaluation. *European Journal of Pharmaceutical Sciences* 99: 310–317.

18 Machado, J.C., Shimizu, F.M., Ortiz, M. et al. (2018). Efficient praziquantel encapsulation into polymer microcapsules and taste masking evaluation using an electronic tongue. *Bulletin of the Chemical Society of Japan* 91: 865–974.

19 Podrazka, M., Baczynska, E., Kundys, M. et al. (2018). Electronic tongue – a tool for all tastes? *Biosensors* 8: 3.

20 Toko, K. (1996). Taste sensor with global selectivity. *Materials Science and Engineering: C* 4: 69–82.

21 Vlasov, Y.G., Legin, A.V., Rudnitskaya, A.M. et al. (1996). Multisensor system with an array of chemical sensors and artificial neural networks (electronic tongue) for quantitative analysis of multicomponent aqueous solutions. *Russian Journal of Applied Chemistry* 69: 848–853.

22 Legin, A., Rudnitskaya, A., Vlasov, Y. et al. (1997). Tasting of beverages using an electronic tongue. *Sensors and Actuators B: Chemical* 44: 291–296.

23 Di Natale, C., Macagnano, A., Davide, F. et al. (1997). Multicomponent analysis on polluted waters by means of an electronic tongue. *Sensors and Actuators B: Chemical* 44: 423–428.

24 Bachmann, T.T. and Schmid, R.D. (1999). A disposable multielectrode biosensor for rapid simultaneous detection of the insecticides paraoxon and carbofuran at high resolution. *Analytica Chimica Acta* 401: 95–103.

25 Tahara, Y. and Toko, K. (2013). Electronic tongues – a review. *IEEE Sensors Journal* 13: 3001–3011.

26 Ha, D., Sun, Q., Su, K. et al. (2015). Recent achievements in electronic tongue and bioelectronic tongue as taste sensors. *Sensors and Actuators B: Chemical* 207: 1136–1146.

27 Lvova, L., Pudi, R., Galloni, P. et al. (2015). Multi-transduction sensing films for electronic tongue applications. *Sensors and Actuators B: Chemical* 207: 1076–1086.

28 Wei, Z., Yang, Y., Wang, J. et al. (2018). The measurement principles, working parameters and configurations of voltammetric electronic tongues and its applications for foodstuff analysis. *Journal of Food Engineering* 217: 75–92.

29 Aono, M. and Ariga, K. (2016). The way to nanoarchitectonics and the way of nanoarchitectonics. *Advanced Materials* 28: 989–992.

30 Ferreira, N.S., Cruz, M.G.N., Gomes, M.T.S.R., and Rudnitskaya, A. (2018). Potentiometric chemical sensors for the detection of paralytic shellfish toxins. *Talanta* 181: 380–384.

31 Manzoli, A., Shimizu, F.M., Mercante, L.A. et al. (2014). Layer-by-layer fabrication of AgCl–PANI hybrid nanocomposite films for electronic tongues. *Physical Chemistry Chemical Physics* 16: 24275–24281.

32 Mercante, L.A., Scagion, V.P., Migliorini, F.L. et al. (2017). Electrospinning-based (bio)sensors for food and agricultural applications: a review. *TrAC, Trends in Analytical Chemistry* 91: 91–103.

33 Wei, Z., Zhang, W., Wang, Y., and Wang, J. (2017). Monitoring the fermentation, post-ripeness and storage processes of set yogurt using voltammetric electronic tongue. *Journal of Food Engineering* 203: 41–52.

34 Fuentes, E., Alcañiz, M., Contat, L. et al. (2017). Influence of potential pulses amplitude sequence in a voltammetric electronic tongue (VET) applied to assess antioxidant capacity in aliso. *Food Chemistry* 224: 233–241.

35 Sobrino-Gregorio, L., Bataller, R., Soto, J., and Escriche, I. (2018). Monitoring honey adulteration with sugar syrups using an automatic pulse voltammetric electronic tongue. *Food Control* 91: 254–260.

36 El Hassani, N.E.-A., Tahri, K., Llobet, E. et al. (2018). Emerging approach for analytical characterization and geographical classification of Moroccan and French honeys by means of a voltammetric electronic tongue. *Food Chemistry* 243: 36–42.

37 Nikolaev, K.G., Ermolenko, Y.E., Offenhäusser, A. et al. (2018). Multisensor systems by electrochemical nanowire assembly for the analysis of aqueous solutions. *Frontiers in Chemistry* 6: 256.

38 de Sá, A.C., Cipri, A., González-Calabuig, A. et al. (2016). Resolution of galactose, glucose, xylose and mannose in sugarcane bagasse employing a voltammetric electronic tongue formed by metals oxy-hydroxide/MWCNT modified electrodes. *Sensors and Actuators B: Chemical* 222: 645–653.

39 Garcia-Hernandez, C., Medina-Plaza, C., Garcia-Cabezon, C. et al. (2018). Monitoring the phenolic ripening of red grapes using a multisensor system based on metal-oxide nanoparticles. *Frontiers in Chemistry* 6: 131.

40 Giménez-Gómez, P., Escudé-Pujol, R., Capdevila, F. et al. (2016). Portable electronic tongue based on microsensors for the analysis of cava wines. *Sensors* 16: 1796.

41 Kilmartin, P.A. (2016). Electrochemistry applied to the analysis of wine: a mini-review. *Electrochemistry Communications* 67: 39–42.

42 Moldes, O.A., Mejuto, J.C., Rial-Otero, R., and Simal-Gandara, J. (2015). A critical review on the applications of artificial neural networks in winemaking technology. *Critical Reviews in Food Science and Nutrition* 57: 2896–2908.

43 Sidel'nikov, A.V., Zil'berg, R.A., Yarkaeva, Y.A. et al. (2015). Voltammetric identification of antiarrhythmic medicines using principal component analysis. *Journal of Analytical Chemistry* 70: 1261–1266.

44 Zil'berg, R.A., Yarkaeva, Y.A., Maksyutova, E.I. et al. (2017). Voltammetric identification of insulin and its analogues using glassy carbon electrodes modified with polyarylenephthalides. *Journal of Analytical Chemistry* 72: 402–409.

45 Aguilar-Lira, G.Y., Gutiérrez-Salgado, J.M., Rojas-Hernández, A. et al. (2017). Artificial neural network for the voltamperometric quantification of diclofenac in presence of other nonsteroidal anti-inflammatory drugs and some commercial excipients. *Journal of Electroanalytical Chemistry* 801: 527–535.

46 Faura, G., González-Calabuig, A., and del Valle, M. (2016). Analysis of amino acid mixtures by voltammetric electronic tongues and artificial neural networks. *Electroanalysis* 28: 1894–1900.

47 Bulbarello, A., Cuenca, M., Schweikert, L. et al. (2012). Hybrid e-tongue for the evaluation of sweetness and bitterness of soft drinks fortified with epigallocatechin gallate. *Electroanalysis* 24: 1989–1994.

48 Zabadaj, M., Ufnalska, I., Chreptowicz, K. et al. (2017). Performance of hybrid electronic tongue and HPLC coupled with chemometric analysis for the monitoring of yeast biotransformation. *Chemometrics and Intelligent Laboratory Systems* 167: 69–77.

49 Gan, Z., Yang, Y., Li, J. et al. (2016). Using sensor and spectral analysis to classify botanical origin and determine adulteration of raw honey. *Journal of Food Engineering* 178: 151–158.

50 Giordano, G.F., Vicentini, M.B.R., Murer, R.C. et al. (2016). Point-of-use electroanalytical platform based on homemade potentiostat and smartphone for multivariate data processing. *Electrochimica Acta* 219: 170–177.

51 Cavanillas, S., Winquist, F., and Eriksson, M. (2015). A self-polishing platinum ring voltammetric sensor and its application to complex media. *Analytica Chimica Acta* 859: 29–36.

52 Paulovich, F.V., Moraes, M.L., Maki, R.M. et al. (2011). Information visualization techniques for sensing and biosensing. *Analyst* 136: 1344–1350.

53 Apetrei, I.M. and Apetrei, C. (2014). Detection of virgin olive oil adulteration using a voltammetric e-tongue. *Computers and Electronics in Agriculture* 108: 148–154.

54 Pigani, L., Vasile Simone, G., Foca, G. et al. (2018). Prediction of parameters related to grape ripening by multivariate calibration of voltammetric signals acquired by an electronic tongue. *Talanta* 178: 178–187.

55 Ceto, X., Saint, C., Chow, C.W.K. et al. (2017). Electrochemical fingerprints of brominated trihaloacetic acids (HAA3) mixtures in water. *Sensors and Actuators B: Chemical* 247: 70–77.

56 Jolliffe, I.T. (1986). Principal component analysis and factor analysis. In: *Principal Component Analaysis*. New York: Springer New York.

57 González-Calabuig, A., Cetó, X., del Valle, M., and Voltammetric Electronic, A. (2018). Tongue for the resolution of ternary nitrophenol mixtures. *Sensors* 18: 216.

58 Bessant, C. and Saini, S. (1999). Simultaneous determination of ethanol, fructose, and glucose at an unmodified platinum electrode using artificial neural networks. *Analytical Chemistry* 71: 2806–2813.

59 Panchuk, V., Lvova, L., Kirsanov, D. et al. (2016). Extending electronic tongue calibration lifetime through mathematical drift correction: case study of microcystin toxicity analysis in waters. *Sensors and Actuators B: Chemical* 237: 962–968.

60 Pérez-Ràfols, C., Serrano, N., Díaz-Cruz, J.M. et al. (2017). A screen-printed voltammetric electronic tongue for the analysis of complex mixtures of metal ions. *Sensors and Actuators B: Chemical* 250: 393–401.

61 Cetó, X., González-Calabuig, A., Crespo, N. et al. (2017). Electronic tongues to assess wine sensory descriptors. *Talanta* 162: 218–224.

62 Tonning, E., Sapelnikova, S., Christensen, J. et al. (2005). Chemometric exploration of an amperometric biosensor array for fast determination of wastewater quality. *Biosensors & Bioelectronics* 21: 608–617.

63 Medina-Plaza, C., García-Hernández, C., de Saja, J.A. et al. (2015). The advantages of disposable screen-printed biosensors in a bioelectronic tongue for the analysis of grapes. *LWT – Food Science and Technology* 62: 940–947.

64 Moraes, M.L., Maki, R.M., Paulovich, F.V. et al. (2010). Strategies to optimize biosensors based on impedance spectroscopy to detect phytic acid using layer-by-layer films. *Analytical Chemistry* 82: 3239–3246.

65 Delezuk, J.A.M., Pavinatto, A., Moraes, M.L. et al. (2017). Silk fibroin organization induced by chitosan in layer-by-layer films: application as a matrix in a biosensor. *Carbohydrate Polymers* 155: 146–151.

66 Paulovich, F.V., Maki, R.M., de Oliveira, M.C.F. et al. (2011). Using multidimensional projection techniques for reaching a high distinguishing ability in biosensing. *Analytical and Bioanalytical Chemistry* 400: 1153–1159.

67 Cipri, A., Schulz, C., Ludwig, R. et al. (2016). A novel bio-electronic tongue using different cellobiose dehydrogenases to resolve mixtures of various sugars and interfering analytes. *Biosensors & Bioelectronics* 79: 515–521.

68 Moraes, M.L., Oliveira, L.P.V., Olivati, C.A. et al. (2012). Detection of glucose and triglycerides using information visualization methods to process impedance spectroscopy data. *Sensors and Actuators B: Chemical* 166–167: 231–238.

69 Gutiérrez, M., Alegret, S., and del Valle, M. (2007). Potentiometric bioelectronic tongue for the analysis of urea and alkaline ions in clinical samples. *Biosensors & Bioelectronics* 22: 2171–2178.

70 Lvova, L., Martinelli, E., Dini, F. et al. (2009). Clinical analysis of human urine by means of potentiometric electronic tongue. *Talanta* 77: 1097–1104.

71 Pascual, L., Campos, I., Vivancos, J.-L. et al. (2016). Detection of prostate cancer using a voltammetric electronic tongue. *Analyst* 141: 4562–4567.

72 Saidi, T., Moufid, M., Zaim, O. et al. (2018). Voltammetric electronic tongue combined with chemometric techniques for direct identification of creatinine level in human urine. *Measurement* 115: 178–184.

73 Huynh, T.-P. and Kutner, W. (2015). Molecularly imprinted polymers as recognition materials for electronic tongues. *Biosensors & Bioelectronics* 74: 856–864.

74 Herrera-Chacon, A., González-Calabuig, A., Campos, I., and del Valle, M. (2018). Bioelectronic tongue using MIP sensors for the resolution of volatile phenolic compounds. *Sensors and Actuators B: Chemical* 258: 665–671.

75 Song, H.S., Jin, H.J., Ahn, S.R. et al. (2014). Bioelectronic tongue using heterodimeric human taste receptor for the discrimination of sweeteners with human-like performance. *ACS Nano* 8 (10): 9781–9789.

76 Liu, Q., Zhang, F., Zhang, D. et al. (2013). Bioelectronic tongue of taste buds on microelectrode array for salt sensing. *Biosensors & Bioelectronics* 40: 115–120.

77 Zhang, W., Chen, P., Zhou, L. et al. (2017). A biomimetic bioelectronic tongue: a switch for on- and off- response of acid sensations. *Biosensors & Bioelectronics* 92: 523–528.

78 Ahn, S.R., An, J.H., Song, H.S. et al. (2016). Duplex bioelectronic tongue for sensing umami and sweet tastes based on human taste receptor nanovesicles. *ACS Nano* 10: 7287–7296.

79 Lee, M., Jung, J.W., Kim, D. et al. (2015). Discrimination of umami tastants using floating electrode-based bioelectronic tongue mimicking insect taste systems. *ACS Nano* 9: 11728–11736.

80 Daikuzono, C.M., Dantas, C.A.R., Volpati, D. et al. (2015). Microfluidic electronic tongue. *Sensors and Actuators B: Chemical* 207: 1129–1135.

81 Braunger, M.L., Shimizu, F.M., Jimenez, M.J.M. et al. (2017). Microfluidic electronic tongue applied to soil analysis. *Chemosensors* 5: 14.

82 Alessio, P., Constantino, C.J.L., Daikuzono, C.M. et al. (2016). Analysis of coffees using electronic tongues. In: *Electronic Noses and Tongues in Food Science* (ed. M.R. Mendez). Amsterdam: Academic Press.

83 Nery, E.W. and Kubota, L.T. (2016). Integrated, paper-based potentiometric electronic tongue for the analysis of beer and wine. *Analytica Chimica Acta* 918: 60–68.

84 Nery, E.W., Guimarães, J.A., and Kubota, L.T. (2015). Paperbased electronic tongue. *Electroanalysis* 27: 2357–2362.

85 Gaál, G., Mendes, M., de Almeida, T.P. et al. (2017). Simplified fabrication of integrated microfluidic devices using fused deposition modeling 3D printing. *Sensors and Actuators B: Chemical* 242: 35–40.

86 Gaál, G., da Silva, T.A., Gaal, V. et al. (2018). 3D printed e-tongue. *Frontiers in Chemistry* 6: 151.

87 Shimizu, F.M., Todão, F.R., Gobbi, A.L. et al. (2017). Functionalization-free microfluidic electronic tongue based on a single response. *ACS Sensors* 2: 1027–1034.

88 Shimizu, F.M., Pasqualeti, A.M., Todão, F.R. et al. (2018). Monitoring the surface chemistry of functionalized nanomaterials with a microfluidic electronic tongue. *ACS Sensors* 2: 716–726.

89 Teixeira, C.A., Giordano, G.F., Beltrame, M.B. et al. (2016). Renewable solid electrodes in microfluidics: recovering the electrochemical activity without treating the surface. *Analytical Chemistry* 88: 11199–11206.

90 Paulovich, F.V., de Oliveira, M.C.F., and Oliveira, O.N. Jr. (2018). A future with ubiquitous sensing and intelligent systems. *ACS Sensors* 3: 1433–1438.

91 Soares, J.C., Shimizu, F.M., Soares, A.C. et al. (2015). Supramolecular control in nanostructured film architectures for detecting breast cancer. *ACS Applied Materials & Interfaces* 7: 11833–11841.

92 Soares, A.C., Soares, J.C., Shimizu, F.M. et al. (2015). Controlled film architectures to detect a biomarker for pancreatic cancer using impedance spectroscopy. *ACS Applied Materials & Interfaces* 7: 25930–25937.

93 Soares, J.C., Soares, A.C., Pereira, P.A.R. et al. (2016). Adsorption according to the Langmuir-Freundlich model is the detection mechanism of the antigen p53 for early diagnosis of cancer. *Physical Chemistry Chemical Physics* 18: 8412–8418.

94 Soares, A.C., Soares, J.C., Shimizu, F.M. et al. (2018). A simple architecture with self-assembled monolayers to build immunosensors for detecting the pancreatic cancer biomarker CA19-9. *Analyst* 143: 3302–3308.

5

Monitoring of Food Spoilage Using Polydiacetylene- and Liposome-Based Sensors

Max Weston, Federico Mazur, and Rona Chandrawati

School of Chemical Engineering and Australian Centre for Nanomedicine (ACN), The University of New South Wales (UNSW Sydney), Sydney, NSW, Australia

5.1 Introduction

The globalization of food trade, intensive agriculture, and an increasingly complex distribution network has expanded the potential for food contamination and spoilage. The United Nations has estimated that up to one-third of all food is lost or wasted [1]. The estimated cost of food safety incidents for the economy of the United States is $7 billion [2]. To reduce food waste and better manage the distribution of food, analytical techniques can be used to monitor contamination and quality. Traditional techniques for contaminant detection include bacterial colony counting [3] and polymerase chain reaction methods [4, 5]. Although specific and sensitive, these techniques are time consuming, laboratory based, and can be expensive for food applications. Nanoscale biosensors are analytical devices that employ a biologically inspired recognition element to selectively detect an analyte. These devices can monitor food storage conditions, quality, shelf life, and contamination. Their portability, speed, ease of operation, and low cost make them an attractive alternate to traditional laboratory-based techniques [6]. Herein, we will discuss recent biosensor developments for the monitoring of food spoilage using polydiacetylene (PDA) and liposome-based systems.

Smart Sensors for Environmental and Medical Applications, First Edition. Edited by Hamida Hallil and Hadi Heidari.
© 2020 The Institute of Electrical and Electronics Engineers, Inc.
Published 2020 by John Wiley & Sons, Inc.

5.2 Polydiacetylene for Visual Detection of Food Spoilage

PDA is formed by the self-assembly of amphiphilic diacetylene (DA) monomers into supramolecular structures such as thin films [7–12] and liposomes [13–22]. Photopolymerization of DA into PDA yields a highly conjugated polymer with unique optical properties [23]. Upon targeted stimulation, PDA undergoes a blue to red and nonfluorescent to fluorescent transition, useful for the construction of colorimetric and fluorescent biosensors (Figure 5.1). Chromatic transition from the blue to red phase can be induced by changes in pH [25, 26], temperature [27–31], mechanical stress [32], and molecular recognition events [11, 20, 33]. Stimulus induces stress on the conjugated backbone shifting absorbance from low energy (blue phase ~ 640 nm) to high energy (red phase ~ 540 nm) [23]. PDA biosensors can be constructed by the conjugation or insertion of recognition elements onto or into the liposome membrane. They are nontoxic and their visible color change make them an attractive low-cost sensor for consumer applications. Consequently, PDA is gaining attention as a colorimetric indicator of food spoilage throughout processing, distribution, and consumption. Recent research has focused on food contaminant detection and the development of freshness indicators to reduce food-related illness, the economic impact of product recalls, and food waste.

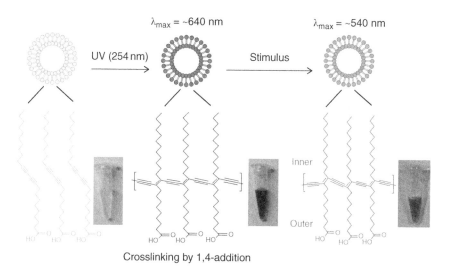

Crosslinking by 1,4-addition

Figure 5.1 Chemical structure of polydiacetylene and its chromatic properties (blue-to-red color transition upon targeted simulation). *Source:* Reproduced with permission from Ref. [24].

5.2.1 Contaminant Detection

Typical food contaminants include bacteria, pesticides, allergens, and toxins. The focus of this section will be on the development of PDA biosensors for pathogenic bacteria detection as it is the major cause of food poisoning, product recalls, and food waste via contamination. Bacteria can be detected using PDA-based sensors directly or indirectly by targeting bacterial cells, metabolites, or other chemical indicators. Direct detection of bacteria using PDA systems draws inspiration from cellular recognition events observed in biology. For example, glycolipids that facilitate cellular recognition of bacteria and viruses in eukaryotic cells can be incorporated into PDA liposomes to serve as recognition elements for pathogens [16–18, 22]. Ma et al. achieved colorimetric detection of *Escherichia coli* by the incorporation of glycolipids DGG and DL3 into the lipid bilayer. They demonstrated that combining glycolipids with shorter chain DA monomers led to systems with higher sensitivity, detecting *E. coli* at 10^8 CFU ml^{-1} [16, 18]. Similarly, Zhang et al. functionalized 10,12-pentacosadynoic acid (PCDA) with the glycolipid MC$_{16}$ to detect *E. coli* [17]. On the other hand, Oliveira et al. used sphingolipids, responsible for signaling and cellular recognition in mammalian cells, by inserting them into PDA vesicles to directly detect *Staphylococcus aureus*, *Salmonella choleraesuis*, and *E. coli* in chicken broth at concentrations as low as 1 CFU ml^{-1} [34].

Indirect detection of bacteria focuses on the recognition of bacterial metabolites and other chemical indicators. A common method is to mimic the structure of biological cell walls by incorporating lipid sections into PDA membranes to detect membrane active compounds excreted by bacteria. Membrane active compounds interact or disrupt the cell membranes of host cells to facilitate the proliferation of bacteria. In sensor design, the compounds disrupt the lipid component within the PDA/lipid structure, inducing a blue to red color change in the PDA backbone. The lipid used can be selected to target bacteria with specificity and to control sensitivity. Scindia et al. developed glass-supported PDA/lipid films that allow visual detection and colorimetric fingerprinting of bacteria [11]. By combining 10,12-tricosadynoic acid (TCDA) with a range of lipids into an arrayed test, they could discriminate between *E. coli* XL1, *E. coli* C600, *Bacillus cereus*, and *Salmonella typhimurium* based on the different profile of membrane active compounds they secrete (Figure 5.2). A characteristic fingerprint is generated for each bacterium based on the reaction of its supernatant with each PDA/lipid test on the array. Kolusheva demonstrated a similar method constructing a PDA/phospholipid liposome array from tricosadiynoic acid and a variety of lipids [35]. The test can discriminate between a range of peptides that are commonly excreted by bacteria.

Alternatively, the headgroup of DA monomers can be functionalized to target specific compounds released by bacteria. Park et al. developed an amine-functionalized PDA liposome that can detect surfactin-producing bacteria [20]. A visible blue to red

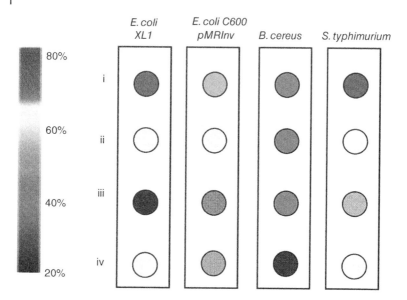

Figure 5.2 Colorimetric bacterial fingerprinting. By combining 10,12-tricosadiynoic acid with a range of lipids into an arrayed test, *E. coli* XL1, *E. coli* C600, *Bacillus cereus*, and *Salmonella typhimurium* could be detected and distinguished based on the different profile of membrane active compounds they secrete. *Source:* Reproduced with permission from Ref. [11].

color change is induced by steric attraction between the positively charged ammonia head group and the negatively charged surfactin at neutral pH (Figure 5.3). Pires et al. similarly functionalized PCDA with N-[(2-tetradecanamide)-ethyl]-ribonamide (TDER) that was capable of detecting *S. aureus* and *E. coli* [36]. The vesicles were placed on a cellulose strip and dried and then used to detect bacteria in water and apple juice. Though capable of detecting the presence of bacteria, typically these systems lack specificity.

More specific systems incorporate aptamers as recognition elements. Aptamers are oligonucleotides that can be engineered to react with binding sites on bacteria or with bacterially excreted compounds with very high specificity. Park et al. functionalized PDA vesicles with aptamers that are complementary to rRNAs of three pathogenic bacteria [37]. The system was immobilized on glass slides and achieved *E. coli*, *Listeria monocytogenes*, and *Salmonella enteritidis* detection between 10^4 and 10^5 CFU ml^{-1}. Similarly, Wu et al. conjugated an aptamer to PDA that specifically binds with lipopolysaccharides on the surface of *E. coli O157:H7* with a detection range of 10^4–10^8 CFU ml^{-1} [38].

Different detection systems are suitable for contaminant detection in different parts of the food life cycle. Low-cost, nonspecific bacteria detection systems such as those constructed from PDA/lipids and PDA with altered head groups are ideal for consumer end applications that may just be concerned with

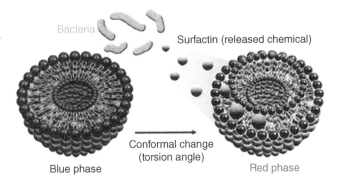

Bacteria

Surfactin (released chemical)

Conformal change
(torsion angle)

Blue phase

Red phase

Figure 5.3 PDA liposomes to detect surfactin-producing bacteria. A visible blue to red color change is induced by steric attraction between the positively charged ammonia head group and the negatively charged surfactin. *Source:* Reproduced with permission from Ref. [20].

overall bacteria count as opposed to type. Applications requiring more specificity in food production, such as culture monitoring in cheese and wine production, may invest more complex aptamer-based sensors. PDA/lipid arrayed systems are attractive as they use low-cost methods to create a sensor that fingerprints bacteria [11], creating specificity that is characteristic of aptamer-based methods.

5.2.2 Freshness Indicators

Traditionally, the freshness of food can be monitored by natural indicators such as smell, taste, and appearance. Although simple, these methods are subjective and unreliable. Date marking tools, such as use-by and sell-by dates, are common instruments that help distributors and consumers manage food products. However, typical date marking tools do not take into account variables that may impact food shelf life such as storage conditions or contamination, and this leads to food waste or the consumption of unsafe food [39–41]. Colorimetric biosensors can be incorporated into intelligent packaging to monitor storage conditions or food constitutions to indicate freshness in real time. These tools enable better management of food and accountability for food quality in all areas of the food life cycle.

Time Temperature Indicators (TTI) are devices that indicate the time–temperature history of a product. This provides information related to the storage and distribution of the food before it is sold or consumed, which provides an indirect indication of the food's quality. Common examples are TTIs on pizza boxes that change color based on the pizza temperature to indicate its freshness. PDA undergoes a blue to red color change over a range of temperatures. The temperature of the color change can be controlled by structural modification [19, 29] or combination with additives [27, 42, 43] making it a versatile building block for TTIs. Typical carboxyl-terminated PDAs such as pentacosadiynoic acid and tricosadiynoic acid undergo a blue to red color transition at temperatures between 50 and 65 °C [19]. The PDA head groups can be modified to increase or decrease the point of temperature change by increasing or decreasing steric attraction between neighboring groups. Park et al. have achieved this by replacing the carboxylic acid head groups of pentacosadiynoic acid with isocyanates to reduce the temperature of color change between 0 and 11 °C [31]. PDA/SiO$_2$ nanocomposites exhibit a similar reduction in its temperature of color change to between 5 and 25 °C as result of ionic interactions between carboxylic acid head groups on PDA and the SiO$_2$ nanoparticles [42, 43]. This system has been applied in the development of TTI for wheat grass [42] and suspended in a PVA film to serve as a TTI for chicken meat [43].

Direct indication of food freshness can be achieved by monitoring its constitution or a substance it releases. Common targets for monitoring a food's freshness include volatile organic compounds (VOCs) [9, 44–47] and substances that are responsible for the off flavors in foods [48]. Nitrogenous food such as meat and fish are highly perishable. Upon spoilage, bacteria metabolize the amino acids in these foods releasing ammonia gas. Park et al. has developed a PDA/PVA film that undergoes a blue to red color change upon exposure to ammonia from spoiling meat at 25 °C [9]. The color transition is attributed to the stripping of a proton from the carboxylic acid head group on the PDA by the ammonia gas, leading to repulsive steric forces between neighboring PDA head groups. Nguyen et al. recently reported a highly sensitive ammonia sensor that can monitor food spoilage at temperatures ranging between −20 °C and room temperature [45] (Figure 5.4). They stabilized PDA vesicles with cellulose nanocrystals in a chitosan

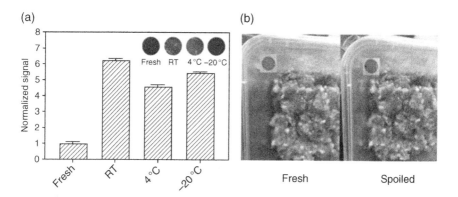

Figure 5.4 (a) Colorimetric response of polydiacetylene films to fresh and spoiled meat at different temperatures (room temperature [RT], 4 °C, and −20 °C) (inset presents the digital images of the sensors). (b) Photographs of polydiacetylene sensors applied on food packaging as an indicator for meat spoilage. *Source:* Reproduced with permission from Ref. [45].

matrix which maximized the PDA concentration on the surface of the film, enhancing the operational temperature, response time, and sensitivity. The film exhibited a blue to red colorimetric response after being exposed to spoiled meat at sub-zero temperatures.

5.2.3 Challenges, Trends, and Industrial Applicability in the Food Industry

The potential applicability of PDA biosensors as contaminant detection devices and freshness indicators has been broadly documented. However, translation of academic study of PDA into commercial application in the food industry has been limited. This is a result of challenges associated with operating in complex food matrices, materials challenges, and consumer resistance.

Food samples can be complicated mixtures of chemicals that may change properties and composition throughout their life. Although this can be exploited to make contaminant detection and food freshness indicators, it is also one of the challenges it must overcome. The ability of PDA to monitor a wide variety of analytes raises issues over specificity. PDA is color responsive to a range of external stimuli (light, impact, temperature, pH, chemical composition, etc.), which can be exposed to the sensor if incorporated into food packaging. This gives potential for false positives, which could exacerbate food waste.

Current work focuses on the incorporation of PDA into materials that maximize specificity while maintaining desired sensor characteristics such as sensitivity, reproducibility, and portability. Recently, PDA has been mixed with a variety of polymers to form films [7, 8, 10, 11] and gels [44, 49] suitable for incorporation into food packaging. Gel and film properties can be controlled by polymer selection and techniques such as crosslinking. Controlling porosity via crosslinking regulates the flow of larger molecules such as proteins that may interfere with sensor performance. Inkjet printing has emerged as a promising technique for the incorporation of PDA sensors onto food packaging [50–52]. PDA sensors can be mixed with clear inks and directly printed onto food items or packaging while maintaining sensor functionality.

In conjunction with the technical challenges of sensing in food samples, consumer acceptance to the pairing of food and technology is unclear and may hinder the incorporation of PDA sensors into food and food packaging. Recent studies suggest that consumers are concerned intelligent packaging may mislead them regarding the product's quality [53] and retailers are concerned that consumers may be pushed to buy only newly displayed items [53]. Furthermore, a clear and unified legislative framework has not been established to govern the use of intelligent packaging with food products [54].

5.3 Liposomes

Liposomes are spherical vesicles, typically 20 nm to 10 μm, which enclose an aqueous environment via single or multiple bilayer membranes composed of phospholipid molecules [55]. Liposome-based assays have been used in food chemistry and microbiology as well as for the detection of harmful elements in food or the human body [56]. They are considered an attractive material in the food industry and are mainly used to improve food flavoring and as nanocarriers of food antimicrobials and metabolites to prevent food spoilage and degradation [57]. An increase in interest toward using liposomes as sensors in the food sector has become apparent, mainly due to their attractive functional properties such as encapsulation capacity, biocompatibility, low cost, and natural composition [6, 58, 59]. The following section aims to provide the current developments in using liposomes for the detection of foodborne pathogens, pesticides, and toxins, concluding with a statement on their stability and their industrial applicability.

5.3.1 Pathogen Detection

5.3.1.1 *Escherichia coli*

E. coli is a common foodborne pathogen which significantly affects human health and the food industry [60]. Illnesses caused by *E. coli* generally consist of a range of conditions including diarrhea, abdominal pain, and headaches, affecting to a greater extent children under 5 years old and older adults [61]. Common sources typically include unpasteurized milk and juices, raw fruits and vegetables, and undercooked meat, among others [62]. Although approaches can be undertaken to prevent infection such as avoiding high-risk foods and proper hygiene, the extensive geographic spread of this bacteria makes controlling food contamination a difficult and unpredictable issue. As such, detection of this pathogen prior to infection is important.

Significant research was conducted over a decade ago using liposomes in different assay formats to detect *E. coli* with sensitivities ranging between 1 and 10^5 CFU ml^{-1} and detection times of five minutes to eight hours [63–69]. More recently, two significant contributions have developed an improved system. Petaccia et al. investigated the effect of cationic liposomes for bacteria detection in drinking water, specifically *E. coli*, *S. aureus*, and *Enterococcus faecalis* [58]. Upon interaction of their liposome system functionalized with surface potential-sensitive fluorophores with each bacterial strain, a detection limit of 10^2 CFU ml^{-1} was achieved within one minute of incubation. To further test their model, they evaluated the fluorescence response when all three strains were present simultaneously. Their results demonstrated a nonadditive fluorescent response, while sensitivities were higher due to the more complex interactions involved. Compared with other nanoparticle-based sensor systems which display higher sensitivities, Petaccia et al. reported a system with shorter analysis times, lower environmental toxicity, ease of preparation, and cost effective synthesis [58]. On the other hand, Bui et al. demonstrated the successful detection of *E. coli*, among other pathogens including *Salmonella* and *Listeria*, in water and food samples [70]. This was achieved by integrating cysteine-loaded liposomes with enzyme-linked immunosorbent assay (ELISA) as well as gold nanoparticles for signal amplification. Visual detection of these pathogens was obtained via a triggered liposome breakdown, releasing the encapsulated cysteine and causing plasmonic gold nanoparticle aggregation, resulting in a red to blue colorimetric shift (Figure 5.5). Using this mechanism, Bui et al. achieved a limit of detection of ~7 attomolar, a six order of magnitude decrease compared to conventional ELISA, while also being the lowest published concentration detected, without using enzymes, by the naked eye [70]. However, this system is only applicable for binary responses (Yes/No), where actual concentration of the analyte is not needed.

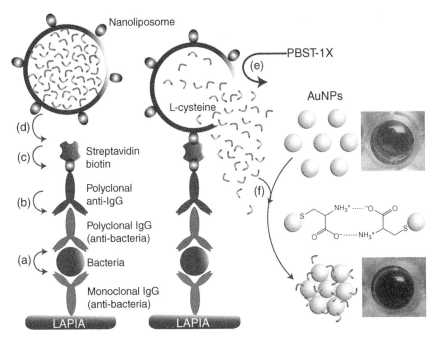

Figure 5.5 Schematic of the liposome-amplified plasmonic immunoassay. Target analytes are captured via sandwich immunoassay. Visual detection of bacteria is obtained via a triggered liposome breakdown, releasing the encapsulated cysteine and causing plasmonic gold nanoparticle aggregation, resulting in a red to blue colorimetric shift. *Source:* Reproduced with permission from Ref. [70].

5.3.1.2 *Salmonella* spp.

In addition to *E. coli*, *Salmonella* is another common foodborne pathogen with typical sources ranging from poultry products to vegetables and generally affects the elderly and children [71, 72]. Common symptoms include diarrhea, abdominal cramps, and vomiting, while infection prevention includes proper hygiene and avoiding high-risk food, especially uncooked food. As with *E. coli*, prior detection is paramount to avoid health issues and prevent infection.

Shin and Kim developed a liposome immunoassay for the detection of *Salmonella* spp. [73]. Their system utilized immunomagnetic separation and immunoliposomes encapsulated with fluorescent dye. After complex formation of the immunomagnetic bead–*Salmonella*–immunoliposome, the quantitative assay was able to achieve detection limits of 27 000 and 5 200 CFU ml^{-1} for *Salmonella enteritidis* and *Salmonella typhimurium*, respectively. Higher sensitivities were obtained when using larger liposomes owing to a greater amount of dye encapsulation. Similarly, Shukla et al. achieved *Salmonella typhimurium* detection using a liposome immunoassay system consisting of anti-Salmonella IgG-tagged immunoliposomes [74]. A limit of detection of 1000 CFU ml^{-1} was obtained, comparably lower than other previously reported techniques, with detection time being significantly reduced due to utilizing immunoliposomes.

In addition, Shukla et al. had also previously developed a liposome-based immunochromatographic strip assay for *Salmonella typhimurium* detection [75]. Liposomes encapsulating dye and conjugated with antibodies were used as immunoliposomes. Their strip assay consisted of a plastic-backed nitrocellulose strip with two antibody zones. Using capillary migration, *Salmonella* concentrated in the antigen capture zone, while unbound liposomes did not, and concentration was measure as a function of color intensity of the capture zone. Using this technique, a detection limit of 100 CFU ml^{-1} was obtained, significantly lower compared to the 10^7 CFU ml^{-1} obtained by commercial immunochromatographic test strip assays that employ gold nanoparticles. Furthermore, their method does not require many washing and incubation steps, in contrast to other immunoassays, making it a rapid and simple approach.

5.3.1.3 Other Bacterium

Inspired by the ELISA analytical technique, Damhorst et al. developed an assay utilizing liposome tagging and ion-release impedance spectroscopy [76]. Liposomes encapsulating ions were conjugated with antibodies and shown to be stable in deionized water, while upon heating became permeable, illustrating a potential toward their use for electrical sensing of antigens. Furthermore, this group showed the ability to quantify viral counts using these liposomes and demonstrated its capacity to detect pathogenic microbes and other biomolecules.

More recently, Shukla et al. [77] and Song et al. [78] successfully detected the bacterium *Cronobacter sakazakii*. Shukla et al. used magnetic nanoparticles coated with anti-*C. sakazakii* and immunoliposomes in both pure cultures and infant formula, achieving a limit of detection of 3300 and 1000 CFU ml^{-1}, respectively [77]. In addition, their assay showed specificity toward *Cronobacter sakazakii* when tested with other pathogens of the same genera. On the other hand, Song et al. employed fluorescent dye encapsulated liposomes conjugated with rabbit anti-*C. sakazakii* IgG, achieving a detection limit of 10^7 CFU ml^{-1} within 15 minutes via an immunochromatographic strip in a specific manner [78]. The significance of these contributions lies in their discussion on multiplexing capabilities towards commercialization, suggesting this limitation in both systems must be overcome prior to validating industrial worth. Nonetheless, since most liposome-based pathogen detection involves fluorescence signals, single-step multiplexing is feasible [77, 78].

5.3.1.4 Viruses, Pesticides, and Toxins

Although mortality rates for bacterial pathogens such as *Salmonella* and *E. coli* are slightly higher, viruses are a significant contributor to outbreaks in food-related illnesses worldwide [79]. The emergence of unknown viral pathogens [80–82] resulting in pandemic outbreaks has already resulted in significant financial expenditure toward their containment. With new viral strains linked to these former ones, their detection has become a priority for health agencies [83]. A contribution by Egashira et al. combined electrochemiluminescence (ECL) with an immunoliposome encapsulating a ruthenium complex [84]. Their detection procedure involved an initial immobilization of hemagglutinin onto a gold electrode, upon which ruthenium-encapsulating immunoliposome would bind via antigen–antibody interactions. The liposome was then lysed with ethanol and the complex was adsorbed onto the gold electrode upon heating. Finally, ECL measurements were taken upon application of a potential. This system was used to detect hemagglutinin molecules in influenza virus with attomole sensitivity. The high sensitivity obtained, significantly higher than with standard ELISA, was attributed to the ECL from the absorbed ruthenium complex onto the gold electrodes.

The extensive use of pesticides, specifically in the food, agriculture, and pest control industries, and their influence on the environment has become an increasingly important issue in recent years. Due to their widespread use, pesticide residues have become pervasive, residing in the soil, atmosphere, groundwater, and agricultural products. This is an issue as even trace amounts of pesticide can lead to respiratory, myocardial, and neuromuscular malfunctioning since they can affect the central nervous system [85]. As such, it is evident that mechanisms to detect pesticides in food and water are needed. Typical analytical techniques include HPLC, GC-MS, and LC-MS/MS, which involve gas chromatography coupled to mass spectroscopy and/or liquid chromatography coupled to tandem mass spectrometry [86]. However, these are time consuming, expensive, require trained personnel, and are not portable. Although techniques to detect pesticides are available, little emphasis has been placed in developing liposome-based sensors for their detection. A unique method was presented by Vamvakaki and Chaniotakis who developed a fluorescence-based nanobiosensor using the enzyme acetylcholinesterase (AchE) for pesticide detection [85]. This enzyme has been shown to improve liposome stability [87] while Vamvakaki et al. [88] had previously demonstrated substrate release improvements when the liposome membrane contained porins. Pyranine, a pH-sensitive fluorescent indicator, was also encapsulated within the liposome to optically transduce enzyme activity. Increasing concentrations of pesticide would result in a decrease in enzymatic activity, thereby decreasing the fluorescent signal produced. As such, Vamvakaki and Chaniotakis biosensor was able to detect dichlorvos and paraoxon, two commonly used pesticides down to a concentration of 0.1 nM [85].

When considering toxin detection, Wen et al. developed a technique for the detection of peanut allergenic proteins in chocolate using Ara h1-tagged liposomes encapsulating fluorescent dye via a lateral flow assay [89]. The signal intensity obtained on the nitrocellulose membrane strips were inversely proportional to the target analyte concentration, achieving a limit of detection of 158 μg of peanuts/g of chocolate within two hours. When compared to published ELISA methods for Ara h1 detection, similar detection limits ranging from 160 to 330 μg of peanuts/g of chocolate were obtained. However, the lateral flow assay presented in Wen et al. [89] contribution benefits from simplicity and lower costs.

Similarly, Ahn-Yoon et al. achieved colorimetric cholera toxin detection using GM1 ganglioside conjugated liposomes [90]. Their sandwich immunoassay employed a nitrocellulose membrane strip upon which GM1-liposomes bound onto the target analyte and were captured by the immobilized antibodies. As with sandwich immunoassay techniques, the color intensity was then measured, obtaining a limit of detection of 10 fg ml^{-1} within 20 minutes. Their assay shows great promise for the ultrasensitive detection of cholera toxin in the field. Using a similar technique Ahn-Yoon et al. also detected botulinum toxin with a limit of detection of 15 pg ml^{-1} within 20 minutes [91].

In contrast, Hirsh et al. reported a technique for nitric oxide (NO) detection via electron paramagnetic resonance [92]. Their unique approach was in combining this common technique for NO detection with liposome-encapsulated spin-trap, which reduces activity inhibition of inducible NO synthase or nitrate reductase. Using this novel approach, a limit of detection of 20 pmol was obtained, showing promise for NO detection in cell lysates and NO-producing tissues.

A gliadin assay developed by Chu and Wen combines the use of anti-gliadin antibody-conjugated immunomagnetic beads, which bind on the gliadin in the sample, with immunoliposomes encapsulating a fluorescent dye which focuses on signal enhancement [93]. In other words, a sandwich complex consisting of the immunomagnetic beads–gliadin–immunoliposomes was used, and a limit of detection of $0.6\,\mu g\,ml^{-1}$ was obtained. More importantly, the developed assay was tested against the standard AOAC-approved ELISA kit, analyzing 20 food samples reliably with no false-negatives. Their study demonstrated high promise toward using alternative techniques for gliadin contamination in the food sector.

5.3.2 Stability of Liposome-Based Sensors

In spite of the aforementioned benefits of liposome, mainly biocompatibility and low cost, one of the main limitations with these systems involves their low mechanical and biochemical stability as well as their low shelf-life during storage. As a result, research is being undertaken to improve these qualities by finding alternate ways of synthesizing these nanoparticles, specifically by changing their composition. In a recent study by Peng et al., hybrid liposomes were prepared using phospholipids and the polysaccharide chitosan, with the intent to improve vesicle stability [94]. They demonstrated greater storage stability compared to phospholipid liposomes, suggesting the enhancement could also improve detection capacities compared to gold nanoparticles, with a potential applicability in the food sector. Alternatively, Virk and Reimhult also attempted to improve liposome stability by creating hybrid vesicles [95]. They combined the biocompatibility inherent in phospholipids with the mechanical stability and tunability of diblock copolymers, creating vesicles with reduced early degradation. These types of systems can improve long-term storage of liposome-based assays, potentially enabling their application in the food sector.

5.3.3 Industrial Applicability of Liposomes

Several examples of the use of liposome nanoparticles as biological sensors have been described, specifically for pathogen, pesticide, and toxin detection. However, several alternative nanomaterials have also been employed as sensors including gold nanoparticles [96], gold nanorods [97], magnetic nanoparticles [98], quantum dots [99], silver nanoparticles [100], and silica nanoparticles [101]. Typical foodborne pathogen or toxin detection takes advantage of optical or electrical properties of nanomaterials [102]; however, liposomes have benefits over other types of nanoparticles mainly due to their biocompatibility and biodegradability. Signal enhancement and transduction is effectively and easily obtained via signal marker encapsulation release or surface functionalization. Moreover, liposomes have shown compatibility with currently available sensing technologies including nanoparticles; immunoassays; and electrochemical, fluorescence, and optical spectroscopy, while demonstrating potential multiplexing capabilities [102].

Liposomes have additional applications in the food industry such as protection of hypersensitive ingredients and preservatives as well as vitamin, enzyme, and antimicrobial encapsulation [103–109]. Greater chemical and environmental stabilities have been demonstrated when encapsulating these materials within liposomes, specifically against temperature, pH, and enzymatic/chemical changes. Despite these benefits, liposomes have also shown cytotoxic effects when using charged liposomes, as demonstrated by Alhajlan et al. [110], as well as trace organic solvent presence depending on the preparation method involved. Furthermore, as was discussed in the previous section, liposome stability is a crucial factor hindering commercialization, which could also be further enhanced when attempting to incorporate into food packaging matrices.

As far as preparation issues are concerned, liposomes suffer from irreproducibility between batches as well as scale-up, sterilization, and instability issues [111]. A variety of preparation methods are available for liposome synthesis including sonication, extrusion, freeze-thawing, microemulsification, reverse phase evaporation, and spray-drying among others [112], each influencing encapsulation and reproducibility to different degrees.

5.4 Conclusions

PDA/liposome-based sensors have been shown to provide a rapid and sensitive way to detect food contaminants and thereby improve food safety. However, several issues hinder transition from lab-scale to commercial-scale use, mainly integration with complex food systems, material challenges, customer resistance, reproducibility, and toxicity. Although significant advancements in this field toward developing more sensitive, specific, and stable assays over the last decade alone were significant, further research into liposome-based sensors is still needed to provide the food industry with efficient and reliable sensors.

References

1 Food and Agriculture Organization of the United Nations. (2011). *Global food losses and food waste - Extent, causes and prevention.*

2 Hussain, M.A. and Dawson, C.O. (2013). Economic impact of food safety outbreaks on food businesses. *Foods* 2: 585–589.

3 Gracias, K.S. and McKillip, J.L. (2004). A review of conventional detection and enumeration methods for pathogenic bacteria in food. *Canadian Journal of Microbiology* 50: 883–890.

4 Candrian, U. (1995). Polymerase chain reaction in food microbiology. *Journal of Microbiological Methods* 23: 89–103.

5 Salihah, N.T., Hossain, M.M., Lubis, H., and Ahmed, M.U. (2016). Trends and advances in food analysis by real-time polymerase chain reaction. *Journal of Food Science and Technology* 53: 2196–2209.

6 Chen, C. and Wang, W. (2015). Liposome-based nanosensors for biological detection. *American Journal of Nano Research and Application* 3: 13–17.

7 Carpick, R.W., Sasaki, D.Y., Marcus, M.S. et al. (2004). Polydiacetylene films: a review of recent investigations into chromogenic transitions and nanomechanical properties. *Journal of Physics: Condensed Matter* 16: R679–R697.

8 Chakarborty, S., Suklabaidya, S., Bhattacharjee, D., and Arshad Hussain, S. (2018). Polydiacetylene (PDA) film: a unique sensing element. *Materials Today: Proceedings* 5: 2367–2372.

9 Park, S., Lee, G.S., Cui, C., and Ahn, D. (2016). Simple detection of food spoilage using polydiacetylene/poly(vinyl alcohol) hybrid films. *Macromolecular Research* 24: 380–384.

10 Ma, B., Fan, Y., Zhang, L. et al. (2003). Direct colorimetric study on the interaction of Escherichia coli with mannose in polydiacetylene Langmuir-Blodgett films. *Colloids and Surfaces B: Biointerfaces* 27: 209–213.

11 Scindia, Y., Silbert, L., Volinsky, R. et al. (2007). Colorimetric detection and fingerprinting of bacteria by glass-supported lipid/polydiacetylene films. *Langmuir* 23: 4682–4687.

12 Cheng, Q. and Stevens, R.C. (1997). Monolayer properties of monosialioganglioside in the mixed diacetylene lipid films on the air/water interface. *Chemistry and Physics of Lipids* 87: 41–53.

13 Weston, M., Kuchel, R.P., Ciftci, M. et al. (2020). A polydiacetylene-based colorimetric sensor as an active use-by date indicator for milk. *Journal of Colloid and Interface Science* doi:10.1016/j.jcis.2020.03.040.

14 Oliveira, C.P.d., Soares, N.d.F.F., Fontes, E.A.F. et al. (2012). Behaviour of polydiacetylene vesicles under different conditions of temperature, pH and chemical components of milk. *Food Chemistry* 135: 1052–1056.

15 Pimsen, R., Khumsri, A., Wacharasindhu, S. et al. (2014). Colorimetric detection of dichlorvos using polydiacetylene vesicles with acetylcholinesterase and cationic surfactants. *Biosensors and Bioelectronics* 62: 8–12.

16 Ma, Z., Li, J., Liu, M. et al. (1998). Colorimetric detection of *Escherichia coli* by polydiacetylene vesicles functionalized with glycolipid. *Journal of the American Chemical Society* 120: 12678–12679.

17 Zhang, Y., Fan, Y., Sun, C. et al. (2005). Functionalized polydiacetylene-glycolipid vesicles interacted with *Escherichia coli* under the TiO_2 colloid. *Colloids and Surfaces B: Biointerfaces* 40: 137–142.

18 Ma, Z., Li, J., Jiang, L. et al. (2000). Influence of the spacer length of glycolipid receptors in polydiacetylene vesicles on the colorimetric detection of *Escherichia coli*. *Langmuir* 16: 7801–7804.

19 Charoenthai, N., Pattanatornchai, T., Wacharasindhu, S. et al. (2011). Roles of head group architecture and side chain length on colorimetric response of polydiacetylene vesicles to temperature, ethanol and pH. *Journal of Colloid and Interface Science* 360: 565–573.

20 Park, J., Ku, S.K., Seo, D. et al. (2016). Label-free bacterial detection using polydiacetylene liposomes. *Chemical Communications* 52: 10346–10349.

21 Seo, D. and Kim, J. (2010). Effect of the molecular size of analytes on polydiacetylene chromism. *Advanced Functional Materials* 20: 1397–1403.

22 Sun, C., Zhang, Y., Fan, Y. et al. (2004). Mannose–*Escherichia coli* interaction in the presence of metal cations studied in vitro by colorimetric polydiacetylene/glycolipid liposomes. *Journal of Inorganic Biochemistry* 98: 925–930.

23 Okada, S., Peng, S., Spevak, W., and Charych, D. (1998). Color and chromism of polydiacetylene vesicles. *Accounts of Chemical Research* 31: 229–239.

24 Weston, M., Tjandra, A.D., and Chandrawati R. (2020). Tuning chromatic response, sensitivity, and specificity of polydiacetylene-based sensors. *Polymer Chemistry* 11: 166–183.

25 Chanakul, A., Traiphol, N., Faisadcha, K., and Traiphol, R. (2014). Dual colorimetric response of polydiacetylene/zinc oxide nanocomposites to low and high pH. *Journal of Colloid and Interface Science* 418: 43–51.

26 Kew, S.J. and Hall, E.A.H. (2006). pH response of carboxy-terminated colorimetric polydiacetylene vesicles. *Analytical Chemistry* 78: 2231–2238.

27 Chanakul, A., Traiphol, N., and Traiphol, R. (2013). Controlling the reversible thermochromism of polydiacetylene/zinc oxide nanocomposites by varying alkyl chain length. *Journal of Colloid and Interface Science* 389: 106–114.

28 Toommee, S., Traiphol, R., and Traiphol, N. (2015). High color stability and reversible thermochromism of polydiacetylene/zinc oxide nanocomposite in various organic solvents and polymer matrices. *Colloids and Surfaces A: Physicochemical and Engineering Aspects* 468: 252–261.

29 Huo, J., Hu, Z., He, G. et al. (2017). High temperature thermochromic polydiacetylenes: design and colorimetric properties. *Applied Surface Science* 423: 951–956.

30 Park, I.S., Park, H.J., Jeong, W. et al. (2016). Low temperature thermochromic polydiacetylenes: design, colorimetric properties, and nanofiber formation. *Macromolecules* 49: 1270–1278.

31 Park, I.S., Park, H.J., and Kim, J.-M. (2013). A soluble, low-temperature thermochromic and chemically reactive polydiacetylene. *ACS Applied Materials & Interfaces* 5: 8805–8812.

32 Hill, S.C., Htet, Y., Kauffman, J. et al. (2013). Polydiacetylene-based smart packaging. In: *Physical Methods in Food Analysis*, 137–154. American Chemical Society.

33 Shriver-Lake, L.C., Taitt, C.R., and Ligler, F.S. (2004). Applications of array biosensor for detection of food allergens. *Journal of AOAC International* 87: 1498–1502.

34 de Oliveira, T.V., Soares, N.d.F.F., de Andrade, N.J. et al. (2015). Application of PCDA/SPH/CHO/Lysine vesicles to detect pathogenic bacteria in chicken. *Food Chemistry* 172: 428–432.

35 Kolusheva, S., Shahal, T., and Jelinek, R. (2000). Peptide–membrane interactions studied by a new phospholipid/polydiacetylene colorimetric vesicle assay. *Biochemistry* 39: 15851–15859.

36 Pires, A.C.d.S., Soares, N.d.F.F., da Silva, L.H.M. et al. (2011). A colorimetric biosensor for the detection of foodborne bacteria. *Sensors and Actuators B: Chemical* 153: 17–23.

37 Park, M.-K., Kim, K.-W., Ahn, D.J., and Oh, M.-K. (2012). Label-free detection of bacterial RNA using polydiacetylene-based biochip. *Biosensors and Bioelectronics* 35: 44–49.

38 Wu, W., Zhang, J., Zheng, M. et al. (2012). An aptamer-based biosensor for colorimetric detection of *Escherichia coli* O157:H7. *PLoS One* 7: e48999.

39 Aschemann-Witzel, J., de Hooge, I., Amani, P. et al. (2015). Consumer-related food waste: causes and potential for action. *Sustainability* 7: 6457–6477.

40 Toma, L., Costa Font, M., and Thompson, B. (2017). Impact of consumers' understanding of date labelling on food waste behaviour. *Operational Research* doi: 10.1007/s12351-017-0352-3.

41 Alexander, P., Brown, C., Arneth, A. et al. (2017). Losses, inefficiencies and waste in the global food system. *Agricultural Systems* 153: 190–200.

42 Suppakul, P., Kim, D.Y., Yang, J.H. et al. (2018). Practical design of a diffusion-type time-temperature indicator with intrinsic low temperature dependency. *Journal of Food Engineering* 223: 22–31.

43 Nopwinyuwong, A., Boonsupthip, W., Pechyen, C., and Suppakul, P. (2013). Formation of Polydiacetylene/Silica Nanocomposite as a Colorimetric Indicator: Effect of Time and Temperature. *Advances in Polymer Technology*, 32: E724–E731.

44 Dolai, S., Bhunia, S.K., Beglaryan, S.S. et al. (2017). Colorimetric polydiacetylene-aerogel detector for volatile organic compounds (VOCs). *ACS Applied Materials & Interfaces* 9: 2891–2898.

45 Nguyen, L.H., Naficy, S., McConchie, R. et al. (2019). Polydiacetylene-based sensors to detect food spoilage at low temperatures. *Journal of Materials Chemistry C* 7: 1919–1926.

46 Jiang, H., Wang, Y., Ye, Q. et al. (2010). Polydiacetylene-based colorimetric sensor microarray for volatile organic compounds. *Sensors and Actuators B Chemical* 143: 789–794.

47 Park, D.-H., Heo, J.-M., Jeong, W. et al. (2018). Smartphone-based VOC sensor using colorimetric polydiacetylenes. *ACS Applied Materials & Interfaces* 10: 5014–5021.

48 Kim, S., Lee, S., Ahn, Y. et al. (2017). A polydiacetylene-based colorimetric chemosensor for malondialdehyde detection: food spoilage indicator. *Journal of Materials Chemistry C* 5: 8553–8558.

49 Fujita, N., Sakamoto, Y., Shirakawa, M. et al. (2007). Polydiacetylene Nanofibers Created in Low-Molecular-Weight Gels by Post Modification: Control of Blue and Red Phases by the Odd−Even Effect in Alkyl Chains. *Journal of the American Chemical Society* 129: 4134–4135.

50 Wu, A., Gu, Y., Stavrou, C. et al. (2014). Inkjet printing colorimetric controllable and reversible poly-PCDA/ZnO composites. *Sensors and Actuators B: Chemical* 203: 320–326.

51 Park, D.H., Park, B.J., and Kim, J.M. (2016). Creation of functional polydiacetylene images on paper using inkjet printing technology. *Macromolecular Research* 24: 943–950.

52 Yoon, B., Shin, H., Kang, E.-M. et al. (2013). Inkjet-compatible single-component polydiacetylene precursors for thermochromic paper sensors. *ACS Applied Materials & Interfaces* 5: 4527–4535.

53 Vanderroost, M., Ragaert, P., Devlieghere, F., and De Meulenaer, B. (2014). Intelligent food packaging: the next generation. *Trends in Food Science and Technology* 39: 47–62.

54 Ghaani, M., Cozzolino, C.A., Castelli, G., and Farris, S. (2016). An overview of the intelligent packaging technologies in the food sector. *Trends in Food Science & Technology* 51: 1–11.

55 Mazur, F., Bally, M., Stadler, B., and Chandrawati, R. (2017). Liposomes and lipid bilayers in biosensors. *Advances in Colloid and Interface Science* 249: 88–99.

56 Shukla, S., Hong, S.Y., Chung, S.H., and Kim, M. (2016). Rapid detection strategies for the global threat of zika virus: current state, new hypotheses, and limitations. *Frontiers in Microbiology* 7: 1685.

57 Emami, S., Azadmard-Damirchi, S., Peighambardoust, S.H. et al. (2016). Liposomes as carrier vehicles for functional compounds in food sector. *Journal of Experimental Nanoscience* 11: 737–759.

58 Petaccia, M., Bombelli, C., Sterbini, F.P. et al. (2017). Liposome-based sensor for the detection of bacteria. *Sensors and Actuators B: Chemical* 248: 247–256.

59 Shukla, S., Haldorai, Y., Hwang, S.K. et al. (2017). Current demands for food-approved liposome nanoparticles in food and safety sector. *Frontiers in Microbiology* 8: 2398.

60 Farahmandfar, M., Moori-Bakhtiari, N., Gooraninezhad, S., and Zarei, M. (2016). Comparison of two methods for detection of *E. coli* O157H7 in unpasteurized milk. *Iranian Journal of Microbiology* 8: 282–287.

61 Rahal, E.A., Kazzi, N., Nassar, F.J., and Matar, G.M. (2012). *Escherichia coli* O157:H7 – clinical aspects and novel treatment approaches. *Frontiers in Cellular and Infection Microbiology* 2: 138.

62 Saxena, T., Kaushik, P., and Mohan, M.K. (2015). Prevalence of *E. coli* O157:H7 in water sources: an overview on associated diseases, outbreaks and detection methods. *Diagnostic Microbiology and Infectious Disease* 82: 249–264.

63 Kim, M., Oh, S., and Durst, R.A. (2003). Detection of *Escherichia coli* O157:147 using combined procedure of immunomagnetic separation and test strip liposome immunoassay. *Journal of Microbiology and Biotechnology* 13: 509–516.

64 Chen, C.S., Baeumner, A.J., and Durst, R.A. (2005). Protein G-liposomal nanovesicles as universal reagents for immunoassays. *Talanta* 67: 205–211.

65 Chen, C.S. and Durst, R.A. (2006). Simultaneous detection of *Escherichia coli* O157:H7, *Salmonella* spp. and *Listeria monocytogenes* with an array-based immunosorbent assay using universal protein G-liposomal nanovesicles. *Talanta* 69: 232–238.

66 Park, S., Oh, S., and Durst, R.A. (2004). Immunoliposomes sandwich fluorometric assay (ILSF) for detection of *Escherichia coli* O157:H7. *Journal of Food Science* 69: M151–M156.

67 DeCory, T.R., Durst, R.A., Zimmerman, S.J. et al. (2005). Development of an immunomagnetic bead-immunoliposome fluorescence assay for rapid detection of *Escherichia coli* O157:H7 in aqueous samples and comparison of the assay with a standard microbiological method. *Applied and Environmental Microbiology* 71: 1856–1864.

68 Ho, J.A.A. and Hsu, H.W. (2003). Procedures for preparing *Escherichia coli* O157:H7 immunoliposome and its application in liposome immunoassay. *Analytical Chemistry* 75: 4330–4334.

69 Ho, J.A., Hsu, H.W., and Huang, M.R. (2004). Liposome-based microcapillary immunosensor for detection of *Escherichia coli* O157:H7. *Analytical Biochemistry* 330: 342–349.

70 Bui, M.P.N., Ahmed, S., and Abbas, A. (2015). Single-digit pathogen and attomolar detection with the naked eye using liposome-amplified plasmonic immunoassay. *Nano Letters* 15: 6239–6246.

71 Siala, M., Barbana, A., Smaoui, S. et al. (2017). Screening and detecting *Salmonella* in different food matrices in southern Tunisia using a combined enrichment/real-time PCR method: correlation with conventional culture method. *Frontiers in Microbiology* 8: 2416.

72 Eng, S.K., Pusparajah, P., Ab Mutalib, N.S. et al. (2015). *Salmonella*: a review on pathogenesis, epidemiology and antibiotic resistance. *Frontiers in Life Science* 8: 284–293.

73 Shin, J. and Kim, M. (2008). Development of liposome immunoassay for *Salmonella* spp. using immunomagnetic separation and immunoliposome. *Journal of Microbiology and Biotechnology* 18: 1689–1694.

74 Shukla, S., Bang, J., Heu, S., and Kim, M. (2012). Development of immunoliposome-based assay for the detection of *Salmonella typhimurium*. *European Food Research and Technology* 234: 53–59.

75 Shukla, S., Leem, H., and Kim, M. (2011). Development of a liposome-based immunochromatographic strip assay for the detection of *Salmonella*. *Analytical and Bioanalytical Chemistry* 401: 2581–2590.

76 Damhorst, G.L., Smith, C.E., Salm, E.M. et al. (2013). A liposome-based ion release impedance sensor for biological detection. *Biomedical Microdevices* 15: 895–905.

77 Shukla, S., Lee, G., Song, X. et al. (2016). Immunoliposome-based immunomagnetic concentration and separation assay for rapid detection of *Cronobacter sakazakii*. *Biosensors & Bioelectronics* 77: 986–994.

78 Song, X., Shukla, S., Lee, G., and Kim, M. (2016). Immunochromatographic strip assay for detection of *Cronobacter sakazakii* in pure culture. *Journal of Microbiology and Biotechnology* 26: 1855–1862.

79 Turcios, R.M., Widdowson, M.A., Sulka, A.C. et al. (2006). Reevaluation of epidemiological criteria for identifying outbreaks of acute gastroenteritis due to norovirus: United States, 1998-2000. *Clinical Infectious Diseases* 42: 964–969.

80 Grinev, A., Daniel, S., Laassri, M. et al. (2008). Microarray-based assay for the detection of genetic variations of structural genes of West Nile virus. *Journal of Virological Methods* 154: 27–40.

81 Bolotin, S., Lombos, E., Yeung, R. et al. (2009). Verification of the Combimatrix influenza detection assay for the detection of influenza A subtype during the 2007–2008 influenza season in Toronto, Canada. *Virology Journal* 6: 37.

82 Tang, Q.H., Zhang, Y.M., Fan, L. et al. (2010). Classic swine fever virus NS2 protein leads to the induction of cell cycle arrest at S-phase and endoplasmic reticulum stress. *Virology Journal* 7: 4.

83 Chen, H., Mammel, M., Kulka, M. et al. (2011). Detection and identification of common food-borne viruses with a tiling microarray. *The Open Virology Journal* 5: 52–59.

84 Egashira, N., Morita, S., Hifumi, E. et al. (2008). Attomole detection of hemagglutinin molecule of influenza virus by combining an electrochemiluminescence sensor with an immunoliposome that encapsulates a Ru complex. *Analytical Chemistry* 80: 4020–4025.

85 Vamvakaki, V. and Chaniotakis, N.A. (2007). Pesticide detection with a liposome-based nano-biosensor. *Biosensors & Bioelectronics* 22: 2848–2853.

86 Grimalt, S. and Dehouck, P. (2016). Review of analytical methods for the determination, of pesticide residues in grapes. *Journal of Chromatography. A* 1433: 1–23.

87 Colletier, J.-P., Chaize, B., Winterhalter, M., and Fournier, D. (2002). Protein encapsulation in liposomes: efficiency depends on interactions between protein and phospholipid bilayer. *BMC Biotechnology* 2: 9.

88 Vamvakaki, V., Fournier, D., and Chaniotakis, N.A. (2005). Fluorescence detection of enzymatic activity within a liposome based nano-biosensor. *Biosensors & Bioelectronics* 21: 384–388.

89 Wen, H.W., Borejsza-Wysocki, W., DeCory, T.R. et al. (2005). A novel extraction method for peanut allergenic proteins in chocolate and their detection by a liposome-based lateral flow assay. *European Food Research and Technology* 221: 564–569.

90 Ahn-Yoon, S., DeCory, T.R., Baeumner, A.J., and Durst, R.A. (2003). Ganglioside-liposome immunoassay for the ultrasensitive detection of cholera toxin. *Analytical Chemistry* 75: 2256–2261.

91 Ahn-Yoon, S., DeCory, T.R., and Durst, R.A. (2004). Ganglioside-liposome immunoassay for the detection of botulinum toxin. *Analytical and Bioanalytical Chemistry* 378: 68–75.

92 Hirsh, D.J., Schieler, B.M., Fomchenko, K.M. et al. (2016). A liposome-encapsulated spin trap for the detection of nitric oxide. *Free Radical Biology & Medicine* 96: 199–210.

93 Chu, P.T. and Wen, H.W. (2013). Sensitive detection and quantification of gliadin contamination in gluten-free food with immunomagnetic beads based liposomal fluorescence immunoassay. *Analytica Chimica Acta* 787: 246–253.

94 Peng, S.F., Zou, L.Q., Liu, W.L. et al. (2017). Hybrid liposomes composed of amphiphilic chitosan and phospholipid: preparation, stability and bioavailability as a carrier for curcumin. *Carbohydrate Polymers* 156: 322–332.

95 Virk, M.M. and Reimhult, E. (2018). Phospholipase A(2)-induced degradation and release from lipid-containing polymersomes. *Langmuir* 34: 395–405.

96 Prasad, D. and Shankaracharya, A.S.V. (2011). Gold nanoparticles-based colorimetric assay for rapid detection of *Salmonella* species in food samples. *World Journal of Microbiology and Biotechnology* 27: 2227–2230.

97 Wang, C. and Irudayaraj, J. (2008). Gold nanorod probes for the detection of multiple pathogens. *Small* 4: 2204–2208.

98 Houhoula, D., Papaparaskevas, J., Zatsou, K. et al. (2017). Magnetic nanoparticle-enhanced PCR for the detection and identification of *Staphylococcus aureus* and *Salmonella enteritidis*. *New Microbiologica* 40: 165–169.

99 Zhu, H.Y., Sikora, U., and Ozcan, A. (2012). Quantum dot enabled detection of *Escherichia coli* using a cell-phone. *Analyst* 137: 2541–2544.

100 Lin, Y.H., Chen, S.H., Chuang, Y.C. et al. (2008). Disposable amperometric immunosensing strips fabricated by Au nanoparticles-modified screen-printed carbon electrodes for the detection of foodborne pathogen *Escherichia coli* O157:H7. *Biosensors & Bioelectronics* 23: 1832–1837.

101 Sun, Q., Zhao, G.Y., and Dou, W.C. (2016). An optical and rapid sandwich immunoassay method for detection of *Salmonella pullorum* and *Salmonella* *gallinarum* based on immune blue silica nanoparticles and magnetic nanoparticles. *Sensors and Actuators B: Chemical* 226: 69–75.

102 Inbaraj, B.S. and Chen, B.H. (2016). Nanomaterial-based sensors for detection of foodborne bacterial pathogens and toxins as well as pork adulteration in meat products. *Journal of Food and Drug Analysis* 24: 15–28.

103 Yang, S.B., Liu, W., Liu, C.M. et al. (2012). Characterization and bioavailability of vitamin C nanoliposomes prepared by film evaporation-dynamic high pressure microfluidization. *Journal of Dispersion Science and Technology* 33: 1608–1614.

104 Wechtersbach, L., Ulrih, N.P., and Cigic, B. (2012). Liposomal stabilization of ascorbic acid in model systems and in food matrices. *LWT- Food Science and Technology* 45: 43–49.

105 Banville, C., Vuillemard, J.C., and Lacroix, C. (2000). Comparison of different methods for fortifying Cheddar cheese with vitamin D. *International Dairy Journal* 10: 375–382.

106 Picon, A., Gaya, P., Medina, M., and Nunez, M. (1995). The effect of liposome-encapsulated bacillus-subtilis neutral proteinase on Manchego cheese ripening. *Journal of Dairy Science* 78: 1238–1247.

107 Rao, D.R., Chawan, C.B., and Veeramachaneni, R. (1995). Liposomal encapsulation of beta-galactosidase – comparison of 2 methods of encapsulation and in-vitro lactose digestibility. *Journal of Food Biochemistry* 18: 239–251.

108 Laridi, R., Kheadr, E.E., Benech, R.O. et al. (2003). Liposome encapsulated nisin Z: optimization, stability and release during milk fermentation. *International Dairy Journal* 13: 325–336.

109 Benech, R.O., Kheadr, E.E., Lacroix, C., and Fliss, I. (2003). Impact of nisin producing culture and liposome-encapsulated nisin on ripening of Lactobacillus added-Cheddar cheese. *Journal of Dairy Science* 86: 1895–1909.

110 Alhajlan, M., Alhariri, M., and Omri, A. (2013). Efficacy and safety of liposomal clarithromycin and its effect on *Pseudomonas aeruginosa* virulence factors. *Antimicrobial Agents and Chemotherapy* 57: 2694–2704.

111 Koynova, R. and Tihova, M. (2010). Nanosized self-emulsifying lipid vesicles of diacylglycerol-PEG lipid conjugates: biophysical characterization and inclusion of lipophilic dietary supplements. *Biochimica et Biophysica Acta-Biomembranes* 1798: 646–653.

112 Akbarzadeh, A., Rezaei-Sadabady, R., Davaran, S. et al. (2013). Liposome: classification, preparation, and applications. *Nanoscale Research Letters* 8: 102.

6

Chemical Sensors Based on Metal Oxides

K. S. Shalini Devi[1], Aadhav Anantharamakrishnan[1,2],
Uma Maheswari Krishnan[1,2,3], and Jatinder Yakhmi[4]

[1] *Centre for Nanotechnology & Advanced Biomaterials, SASTRA Deemed University, Thanjavur, India*
[2] *School of Chemical & Biotechnology, SASTRA Deemed University, Thanjavur, India*
[3] *School of Arts, Science & Humanities, SASTRA Deemed University, Thanjavur, India*
[4] *Bhabha Atomic Research Centre, Mumbai, India*

6.1 Introduction

Qualitative and quantitative analysis of molecules of interest is of paramount importance in healthcare, process, food, and pharmaceutical industries, as well as in environmental monitoring. The device that enables the detection of the molecule of interest or the analyte is known as a sensor. The chemical information transformed by the sensing element of the sensor, which can be a particular ion or element or single chemical or biological molecule, into an analytically useful signal is termed a "Chemical sensor" [1]. A wide range of transduction methods have been reported for chemical sensors. These include electrical, optical, electrochemical, piezoelectric, surface acoustic waves, microcantilevers, etc. The change in electrical properties has been found to be an extremely versatile strategy for detection of gaseous analytes. Generally, the chemical sensors have been employed for detection of volatile molecules in the gas phase while cantilever and mass-based changes have been found to be efficient in liquids.

Metal oxides (MOx) have elicited interest for gas sensing applications due to their affordability, significant change in conductivity on exposure to the target molecule, tunable properties by doping, and the ability to be easily interfaced with different transducing systems [2]. MOx sensors exhibit promise for detecting toxic pollutants (CO, H_2S, NO_x, SO_2, volatile organic compounds [VOCs], etc.),

Smart Sensors for Environmental and Medical Applications, First Edition. Edited by
Hamida Hallil and Hadi Heidari.
© 2020 The Institute of Electrical and Electronics Engineers, Inc.
Published 2020 by John Wiley & Sons, Inc.

explosive gases (H_2, CH_4, flammable organic vapors, etc.), and markers in exhaled breath when compared to other types of organic sensors [3]. As most applications of MOx sensors are based on gaseous markers, it is considered a mimic of the human olfactory system that can detect and recognize over several thousand different odors. Hence, MOx sensors are often referred to as the electronic nose (e-Nose) [4].

6.2 Classes of MOx-Based Chemical Sensors

MOx sensors have been classified as transition MOx (Fe_2O_3, WO_3, V_2O_5, CuO, NiO, etc.) and nontransition MOx sensors based on the electronic structure of MOx employed [5]. The nontransition MOx sensors include the pretransition series such as MgO and Al_2O_3 or post-transition type like In_2O_3 and SnO_2 [6]. The pretransition MOx possess high stability but poor electronic properties and hence have been generally less preferred for use as conductive gas sensors [5]. The transition MOx sensors possess excellent conducting properties and variable chemical valency that enable sensitive detection of even trace quantities of the analyte but may sometimes have stability issues. Based on the dimensions of the sensing element, MOx sensors can be classed as bulk MOx sensors and nanosensors that have dimensions in the nanometer range. The use of nano-dimensional structures give a higher surface area-to-volume ratio that improves the sensing efficiency when compared to bulk sensors. MOx sensors can also be categorized based on the sensing mechanism as conductometric, surface-ionization, photoluminescent, gravimetric, thermometric, and magnetic sensors [7]. Among these types, the conductometric sensors have been extensively reported in literature due to their suitability for analysis of gaseous analytes. The semiconductor MOx sensors have been classified as n-type (SnO_2, ZnO, TiO_2, WO_3, Ga_2O_3, Fe_2O_3, CeO_2, In_2O_3, etc.) and p-type MOx sensors (Cr_2O_3, NiO, Co_3O_4, V_2O_5, CuO) based on the type of charge carriers. While the n-type sensors have electrons as the charge carriers, the p-type involves holes as the major charge carriers. Based on the phase of analysis, the chemical sensors have also been classified as gas sensors, and liquid and solid particulate sensors [8].

6.3 Synthesis of MOx Structures

A host of methods have been employed for the synthesis of MOx particles with well-defined dimensions and morphology. These include hydrothermal, carbothermal reduction, ultrasonic irradiation, electrospinning, anodization, sol-gel, molecular beam epitaxy, chemical vapor deposition, molten-salt, solid-state chemical reaction, thermal evaporation, vapor-phase transport, RF sputtering, aerosol, gas-phase-assisted nanocarving, UV lithography, and dry plasma etching. Among these, the sol-gel process and hydrothermal methods, both of which are bottom up approaches, and sputtering techniques which form part of top-down methods are extensively employed for synthesis of MOx. The method of synthesis has a profound influence on the shape, size, and sensing characteristics of the MOx [9]. In recent years, one-dimensional structures such as wires and fibers are also garnering attention for sensing applications due to their increased surface area that auger well for sensing of analytes [10]. Electrospinning has emerged as an elegant method to fabricate MOx nanofibers. However, scaling up remains an issue with electrospinning. Molecular imprinting and template-based synthesis methods have also been recently explored for obtaining highly ordered MOx structures with nearly monodisperse population [11].

6.4 Mechanism of Sensing by MOx

MOx have been mostly employed as chemiresistors where their resistance is altered on exposure to the target molecule [12]. The chemiresistors exhibit both physisorption and chemisorption. The first step involves the adsorption of oxygen molecules on the surface of the MOx. The oxygen abstracts electrons from the n-type semiconductor thereby resulting in the formation of anionic moieties such as O_2^-, O^{2-}, and O^-. This depletes the charge carrier concentration at the surface resulting in an increase in resistance.

$$O_2\left(gas\right) + e^- \rightarrow O_2\left(adsorbed\right)$$

$$O_2\left(adsorbed\right) + e^- \rightarrow O_2^-$$

$$O_2^- + e^- \rightarrow 2O^-$$

The grain size, chemical nature, and pores on the MOx layer influence the chemisorption of oxygen. When a reducing gas like carbon monoxide is introduced, the resistance will drop due to the interaction of the gas with the oxygen layer. In the case of an oxidizing gas such as NO_2, the resistance will decrease as a result of further depletion of the charge carriers. The oxygen atoms extracted from the surface of reducing gases will behave as donors on the n-type MOx, whereas the oxygen atoms present in the surface of oxidizing gases act as acceptors. The involvement of the reducing gases tends to increase conductance and vice versa for the oxidizing gases [7]. The sensor response in the presence of reducing and oxidizing gases will be reversed for p-type MOx semiconductors. Physisorption of analyte gases on the surface of the MOx occurs through weak van der Waal's interactions thereby modifying the electronic properties of the receptor surface. Most MOx sensors are limited by a lack of specificity toward a target molecule. However, the specificity can be enhanced by altering the stoichiometry, porosity, and grain size of the MOx particles as well as introduction of appropriate dopants.

In the surface ionization mechanism, when MOx are heated in the presence of an applied electric field, it leads to ionization of the surface-adsorbed molecular layer. As a result, there is an increase in the conductance which is recorded for quantification of the target molecules. MOx nanowires have been found to be effective in quantification of the gaseous molecules by the surface ionization method [13]. The only drawback of this method is that the presence of moisture interferes with the sensor response. In addition, changes in the photoluminescence properties of MOx such as ZnO have also been explored as a strategy to quantify the target molecules. The quenching of the photoluminescence by the target molecule in a concentration-dependent manner was used to quantify the analyte. The quenching could be due to changes in the thickness of the space charge region due to the adsorbed analyte layer or a consequence of the inhibition of radiative transfer process by the adsorbed analyte [14]. Few attempts to quantify the changes in the magnetization values of the MOx layer in a known applied magnetic field postexposure to the analyte have also been reported in literature [15]. However, the most extensively employed mechanism for gas sensing is the chemoresistive method owing to its simple instrumentation setup, rapid response, and portability [16].

6.5 Factors Influencing Sensing Performance

The performance of MOx-based chemical sensors is influenced by many factors that can be broadly categorized as chemical, structural, and environmental parameters.

The chemical composition and surface modifications of the MOx sensor are key factors that affect the sensing performance. The blending of more than one MOx has been shown to have a positive impact on the sensing parameters for several analytes. For instance, when a blend of tin oxide and zinc oxide was employed for detection of butyraldehyde, it was found that while tin oxide dehydrogenated the analyte, zinc oxide catalyzed the decomposition of the analyte. Together the catalytic activities of the two MOx in the blend synergistically contributed to improved sensitivity toward the detection of butyraldehyde [5]. Another possibility that arises when a mixture of oxides are employed is the formation of heterojunctions whose conductance properties get altered in the presence of a target molecule. However, it is important to note that the percentage composition of the MOx in the blend can influence the sensor response [17]. However, a few combinations such as zinc oxide and indium oxide displayed an antagonistic effect and hence the choice of the MOx combination and composition both become key parameters for designing an efficient sensor [5]. Similar to the blending of MOx, introduction of small quantities of metal atoms in the MOx layer known as "doping" has also been found to be beneficial for sensing. The metal centers act as catalytic sites that enhance the reactions at the MOx surface. In a typical example, palladium-doped SnO_2 sensors were employed for quantification of hydrogen gas where the palladium served as a catalyst that enhanced the sensor response [18].

The morphology and dimensions of the MOx sensing element play an important role in determining the sensitivity. Several nanostructures such as spheres, rods, flakes, flowers, and fibers [19] have been explored for sensing and the results demonstrate that the morphology of the structures influence the sensing characteristics. SnO_2 nanofibers fabricated by electrospinning and nanowires formed by thermal-evaporative conduction (TEC) were compared for their gas detection efficiency [20]. The electrospun nanofibers exhibited superior sensing performance when compared to their TEC counterparts. The performance was further improved in the presence of palladium nanoparticles. Better results were obtained at higher temperatures for the nanowires unlike the nanofibers where a decrease in response was recorded. This difference in sensing performance was attributed to the higher density of chemisorption sites at low temperatures and subsequent loss of chemisorbed oxygen at higher temperatures in the nanofibers. The shape of the MOx nanocrystals also influences its sensing characteristics. A polyhedral structure when compared to an octahedron of the same MOx was found to display superior sensing performance which may be attributed to its high surface area-to-volume ratio [5]. In another report, two different sizes of ZnO were employed for detection of isoprene. The nanoparticles about 5 nm in size exhibited better sensitivity and faster response when compared to the 25 nm particles due to the higher band gap and surface area-to-volume ratio of the smaller nanostructures [21]. Thus, it is evident that both size and shape of the MOx particles exert a strong influence on the sensing characteristics. Many researchers have highlighted the relation between the grain size and the thickness of the space charge region formed by the depletion of electrons by the adsorbed oxygen molecules on the surface of the MOx (Figure 6.1). If the grain size is much larger than the thickness of the space charge layer, then the conductivity of the MOx layer is independent of the surface reaction and it remains dependent on the number of inner charge carriers. This phenomenon is denoted as the boundary controlling model. If the thickness of the space charge layer is equal to or slightly smaller than the grain size, the conductivity depends on the charges at the interface between two grains known as the "neck"; the regime is denoted as the neck controlling model. In the third case known as the grain controlling model, the grain size is lesser than the thickness of the space charge layer, and the conductivity is dependent only on the surface reaction as there are no inner charge carriers. This is ideal for highly sensitive detection of an

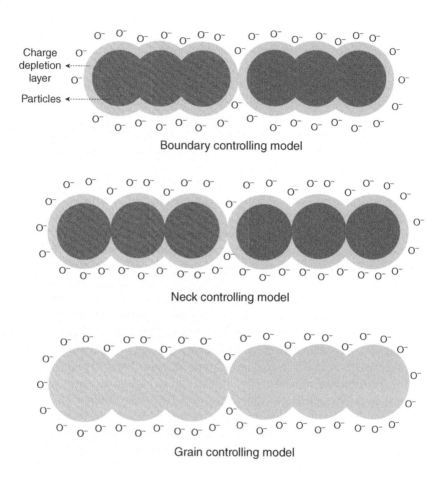

Figure 6.1 Schematic representation of different models correlating grain size with the thickness of the space charge region.

analyte as even lesser number of molecules interacting with the surface can produce a significant change in the conductivity of the MOx layer [22].

Humidity and moisture content in the environment have a negative impact on the sensing capabilities of a MOx sensor. The water molecules tend to chemisorb on the MOx surface leading to formation of the hydronium and hydroxyl ions that are retained on the surface through electrostatic interactions with the metal cations [23]. The hydronium ion transfers its proton to an O^{2-} ion in the vicinity thereby forming OH^-. This process not only depletes the surface oxygen but decreases the resistance of the MOx layer. The chemisorbed layer decreases the electron donation tendency, which in turn reduces the sensitivity of the MOx surface toward other gaseous analytes. The chemisorbed layer promotes further physisorption of additional layers of water molecules which are not very tightly associated with the surface. The strong electrostatic environment of the chemisorbed layer facilitates the dissociation of the physisorbed water molecules to form ions. The hydronium ion thus formed can undergo the Grotthuss chain reaction where it transfers its proton to a neighboring water molecule, which in turn donates the proton to an adjacent water molecule and so on [24]. The transfer of proton to O^{2-} or H_2O depends on the surface coverage. The adsorbed water molecules further restrict the interaction of the analyte gases with the MOx surface. In order to reduce the number of physisorbed water molecules, most MOx sensors are operated at elevated temperatures to remove the physisorbed water layer. The chemisorbed hydroxyl ions are generally removed at temperatures exceeding 450 °C, thereby further increasing the operating temperature of the sensor. This property has also been exploited for the development of MOx-based humidity sensors [25].

Temperature itself has an important role in determining the efficiency of detection. At low temperatures, the detection efficiency of MOx sensors is poor owing to slow kinetics of interaction [5]. The efficiency progressively increases with increase in operating temperatures and peaks at a specific value that is unique for the type of gaseous analyte and MOx being employed. Further increase in temperature decreases the sensor efficiency due to desorption of the gaseous analyte from the MOx surface. Thus, it is essential to ensure that the operating temperature of every MOx sensor does not exceed the working range. Apart from efforts to reduce the grain size of the MOx, use of ZnO and indium oxide structures that spontaneously react with specific gases like H_2S to form the respective sulfides at room temperature contributes to the sensing of these gases at ambient conditions [26].

6.6 Applications of MOx-Based Chemical Sensors

MOx-based chemical sensors have been extensively explored for numerous applications spanning different sectors. The advances in materials science and nanotechnology continue to introduce novel combinations of MOx for ultrasensitive detection of analytes that have contributed to new vistas for analysis (Figure 6.2).

The following sections highlight some remarkable uses of MOx-based chemical sensors in the recent decades.

6.6.1 MOx Sensors for Environmental Monitoring

Noxious effects of gases like sulfurous oxides (SOx), carbon dioxide (CO_2), nitrous oxides (NOx: NO, NO_2, N_2O, N_2O_3, N_2O_5), hydrogen sulfide (H_2S), carbon monoxide (CO), hydrocarbons, NH_3, VOCs etc., cause severe threat not only to humans

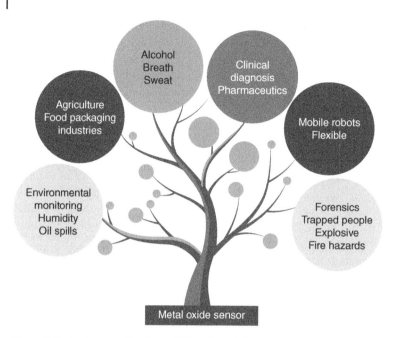

Figure 6.2 Various applications of MOxs-based chemical sensors.

but also to other ecosystems [27]. Several MOx have been explored for the detection of various pollutant gases. SnO_2 sensor is among the most extensively investigated for a wide range of gaseous analytes like NO [28], NO_2 [29], and NH_3 [30]. Fe-doped WO_3 mesoporous hollow nanospheres displayed a sensing range of 10–1000 ppb toward NO_2 at 100–140 °C [31] emphasizing the catalytic role of the dopant. ZnO nanoparticles is another extensively investigated MOx that has been used in several configurations like platelets, rings, combs, rods, etc., for the detection of flammable gases. Functionalized ZnO nanorods along with AlGaN/GaN heterostructures have been investigated for monitoring of NH_3 in air. The sensitivity of the electrode was 2 ppm at 5 V [32]. Ba^{2+}-doped amorphous TiO_2 hollow spheres prepared by template method under irradiation were tuned as photoelectrochemical sensor for the detection of sulfur dioxide (SO_2). The fabricated sensor demonstrated excellent stability and was used to assess SO_2 levels in industrial effluents. The detection range of the sensor was between 1 and 4000 pmol l^{-1} and the LoD was 0.4 pmol l^{-1}. The current paradigm in environment monitoring is to employ a panel of MOx sensors to detect different gases in a mixture rather than individual gases. Real-time and remote monitoring of gaseous pollutants are also under active development with attempts to enable optimum sensor performance at ambient conditions. Table 6.1 highlights the salient features of a few MOx-based sensors for detection of pollutants that have been documented in recent years.

Table 6.1 Sensing parameters of MOx-based sensors used for detection of gaseous pollutants.

Modifier	LOD	Linear range	Response time (sec)	Gas	Reference
CuO nanowires (FET)	5 ppm	50–800 ppm	<10	CO	[33]
Al-doped ZnO	<750 ppb	5–50 ppm	6–8	CO	[34]
ZnO nanowires/Au nanoparticles	—	5–100 ppm	40	CO	[35]
SnO_2/RGO composite	2 ppm		4–12	NO_2	[36]
ZnO nanorods	10 ppm	10–140 ppm	4	H_2S	[37]
ZnO thin film	—	5–100 ppm	20	NH_3	[38]
CZO nanoflowers	10 ppm	10–100 ppm	32.3	NH_3	[39]
V_2O_5/PVAC composites	100 ppb	0.8–8.5 ppm	50	NH_3	[40]
RGO/CO_3O_4 composite	—	5–100 ppm	4	NH_3	[41]
ZnO nanopencils	5 nM	15 nM–0.5 mM	<10	NH_3	[42]
SnO_2/Ppy	—	10–100 ppm	3–4	NH_3	[43]
CuO decorated Gr hybrid nanocomposite	—	0.25–100 ppm	70–76	CO	[44]
$LaFeO_3$ nanocrystalline powder	—	500–2000 ppm	4 min	CO_2	[45]
ZnO/MWCNT	40 ppm	40–200 ppm	8–23	CO	[46]
Reduced GrO-WO_3 nanocomposite	0.5 ppm	5–20 ppm	9 min	NO_2	[47]
Fe_2O_3 thin film	—	10–200 ppm	12	NO_2	[48]
Ppy-WO_3 hybrid nanocomposite	5 ppm	5–100 ppm	—	NO_2	[49]
ZnO hierarchial nanostructures	1 ppm	1–100 ppm	50	NO_2	[50]
Nano-$BiFeO_3$	5 ppm	—	20	SO_2	[51]
(i) RGO/SnO_2	10 ppm	10–500 ppm	2.4 min	SO_2	[52]
(ii) MWCNT/SnO_2			5.3 min		
Nano crystalline $BaSnO_3$	10 ppm	10–40 ppm	—	SO_2	[53]
LaCaFeO	3 ppm	1–10 ppm	86 s	SO_2	[54]
WO_3 nanoplates	5 ppm	1–100 ppm	3 min	NO_2	[55]
Ag/mesoporous WO_3	100 ppb	0.1–1 ppm	5.05 min	NO_2	[56]
Poly crystalline CuO	sub ppm	0.5–10 ppm	—	H_2S	[57]

6.6.2 MOx Sensors in Clinical Diagnosis

The detection of volatile gaseous markers from exhaled breath to diagnose a disease is a rapidly growing field and has given rise to a new area of diagnosis known as "breath printing" [21]. It is pertinent to note that in many cases, a single marker may not truly reflect the health condition of an individual and a panel of markers may be required for identifying the risk. Attempts are underway to identify novel markers for specific diseases that would serve as the basis for development of a panel of sensors comprising an e-nose (Figure 6.3).

Acetone, a biomarker for both type-I and type-II diabetes was detected using C-doped WO_3 synthesized through facile calcination-assisted template method. The remarkable sensitivity of the sensor was attributed to the carbon doping while the long-term stability was ascribed to the lower operating temperature (300 °C) which is below the C-doped WO_3 crystal forming temperature. Diabetic patient samples tested elicited a response of greater than 1.8 ppm in this sensor [58]. Estimation of toluene in the exhaled air of people is considered to be a reliable indicator for lung cancer which ranks among the most aggressive cancers. A toluene sensor employing WO_3 nanofibers decorated with Pd catalyst impregnated in the pores of WO_3 fibers as the sensing element was reported recently. The observed sensing parameters such as response of 5.5 (R_{air}/R_{gas}), response time of

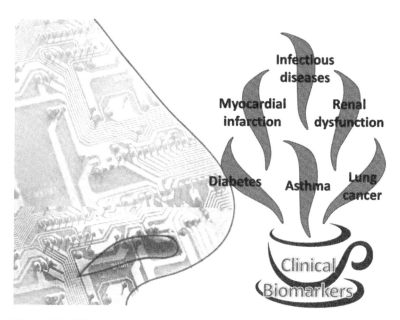

Figure 6.3 Schematic representation of an e-nose for sniffing volatile markers that are hallmarks of diseases affecting mankind.

10.9 seconds, and a recovery time of 16.1 seconds at 1 ppm, can be attributed to the homogenous distribution of Pd nanoparticles on WO_3 and grain size reduction occurring due to Pd-doping in WO_3 [59].

MOx-based chemiresistive electronic nose (CEN) with Au-functionalized villi-like SnO_2 and WO_3 nanostructures (Au-VLNs) was engineered to quantify NO and NH_3 levels in exhaled air which are potent indicators of asthma and renal dysfunction. The sensor exhibited a sensitivity of 1 ppm for NO and 20–10 ppm for NH_3 vapors at extreme humidity due to large surface area, porosity of the VLNs, and efficient potential barriers arising due to nano-necks between individual nanostructures [60]. To monitor β-hematin, a biomarker for malaria, Au-MOx-modified electrodes have been fabricated with improved electron transfer characteristics. Cyclic voltammetric studies on these electrodes revealed a reduction peak at −0.72 V corresponding to β-hematin as a result of formation of Fe(II). For $Au-Fe_2O_3$ and $Au-Al_2O_3$ electrodes, the detection was a consequence of surface passivation postcatalysis impeding electron migration [61]. Table 6.2 lists some recent reports on MOx-based sensors that have been employed for detection of VOCs from breath.

Recently, nanostructured plate-like copper oxide films prepared by successive ionic layer adsorption and reaction (SILAR) and doped with zinc were employed for monitoring hydration levels in a range of sweat concentrations. Upon Zn doping, substitution of zinc ions at copper sites released two electrons that elevated the charge carrier concentration and altered the resistivity of the CuO film [88]. A sensor for simultaneous determination of dopamine, uric acid, and ascorbic acid was fabricated using iron oxide-carbon nanotube composite [89]. The multianalyte sensor combined the advantages of the sensing abilities of iron oxide and the fast electron transfer properties of the nanostructured carbon nanotubes. MOx have been extensively used as immobilization matrices for enzymes and antibodies that serve as the sensing element for biologically relevant molecules [90]. This area has gained much importance in the recent decades and has evolved into an independent field of biosensors.

6.6.3 MOx Sensors in Pharmaceutical Analysis

Several MOx-based sensors have been employed as electronic tongue (e-tongue) in hospitals to detect trace quantities of drug molecules or their metabolites in body fluids as well as in pharmaceutical industries where these sensors could be deployed for quality control of the formulations. A 1D gelatin-infused CuO nanoneedle-based sensor was fabricated for detection of an anti-tuberculosis drug, Rifampicin, a competitive inhibitor of *Mycobacterium* RNA Polymerase. The sensor exhibited a LoD of 9.4 nM. The superior performance of the sensor was ascribed to the chemical reaction between the hydroxyl groups in rifampicin's

Table 6.2 Salient features of MOx sensors employed for detection of volatile organic compounds.

Working electrode	Response time (s)	Recovery time (s)	Limit of detection	Temperature (°C)	VOC identified	Reference
C-doped WO$_3$	3–9	6–12	0.2 ppm	300	Acetone	[58]
ZnO microcrystals	1.5	3		300	Acetone	[62]
ZnO/ZSM-5	105	185	10 ppm	90	Ethanol	[63]
TiO$_2$/ZSM-5	10	200		160	Ethanol	[63]
Ni-doped SnO$_2$ microstructures	14	13	0.09 ppm	260	Formaldehyde	[64]
	17	14	0.17 ppm		ethanol	
Cr-ITO thin film sensor	38	53	30 ppm	756	Benzene	[65]
CB coadsorped ZnO nanocomposite	13	—	18.75 pM	—	Benzaldehyde	[66]
ZnO flowers	53	151	100 ppm	350	Toluene	[67]
Coralloid SnO$_2$ nanostructures	20	15	50 ppm	180	Benzaldehyde	[68]
					Acetone	
Au-functionalized ZnO	36	45	100 ppm	340	Toluene	[69]
Sn-doped NiO	298	223	0.3 ppm	225	Xylene	[70]
Pt-loaded Al$_2$O$_3$ filter/WO$_3$	—	—	1 ppm	300	Benzene	[71]
Ni-doped ZnO nanowires	15	—	5 ppm	400	p-Xylene	[72]
Double layered thin film/Ni deposited porous alumina	20	300	0.07 ppm	340	Xylene	[73]
Au-loaded WO$_3$-H$_2$O nanocubes	—	—	200 ppb	255	Xylene	[74]

Ti-doped ZnO	60	300	5 ppb	325	Isoprene	[75]
ZnO quantum dots	65	145	1 ppm	350	Isoprene	[76]
Cu-doped ZnO nanopowders	—	—	0.1 ppm	300	Propane	[77]
ZnO hollow spheres	36	9	10 ppm	385	Butanol	[78]
Ordered mesoporous In-(TiO_2/WO_3) nanohybrid	1.5	1.5	1 ppm	200	Butanol	[79]
PtO_2 nanoparticles CuO polyhedrons	2.4	9.2	1 ppm	180	Butanol	[80]
Au-loaded mesoporous WO_3	10	35	10 ppm	250	Butanol	[81]
SnO_2/Al_2O_3/Nb_2O_5/SiO_2	3	600	1 ppm	300	Acetonitrile	[82]
RGO/SnO_2 p-n heterojunction aerogels	2.43	1.06	5 ppb	600	Phenol	[83]
Cr_2O_3 decorated nanotubes	4	—	0.64 nM	25	Phenol	[84]
WO_3 thin films	—	—	10 ppm	846	Dodecane	[85]
Nanostructured WO_3 films	—	—	1 ppm	300	Dodecane	[86]
Au/TiO_2 Pecan Kernel	4	5	100 ppm	375	Toluene	[87]

naphthol ring and the anionic CuO nanoneedles through the C–O group in gelatin [91]. The anticancer drugs doxorubicin and dasatinib used widely in the treatment of breast cancer have been detected using the ZnO/ionic liquid/CPE modification in blood and serum samples. The sensor exhibited a detection limit of 9.0 and 0.5 nM, respectively, for doxorubicin and dasatinib and remained unaffected by the presence of several other biological molecules like glucose, methionine, valine, etc. [92]. WO_3 in combination with graphene adsorbed over a glassy carbon electrode (GCE) facilitated the detection of ranolazine, an antianginal drug, at concentrations as low as 0.13 µM [93]. These studies have highlighted that MOx-based composites could be invaluable in pharmaceutical sector for interference-free quantification of drug molecules, intermediates, as well as drug metabolites.

6.6.4 MOx-Based Sensors in Food Analysis

Several MOx-based chemical sensors have been reported with excellent sensitivity and detection range for quality assessment of fruits, vegetables, and other food products and also for evaluating their edibility. Ascorbic acid was determined in orange, lemon, apple juices, vegetables, and food supplements using a ZnO/CNT nanocomposite deposited over a 1-methyl-3-butylimidazolium bromide bonded CPE. The sensor was able to detect ascorbic acid up to a concentration of 0.07 µM as a result of the improved electro-oxidation of ascorbic acid at the composite electrode surface [94]. Since microbial contamination is ranked as a major propagator of food spoilage, an electronic nose comprising SnO_2, MoO_3, Mo-doped SnO_2, and tin oxide was developed to discern the volatile fingerprint of *Enterobacteriaceae* in mixed vegetable soups. This group demonstrated the significance of incubation time that facilitates the recognition of a specific VOC among a mixture of emitted gases [95]. A MOx gas sensor array comprising a blend of tin and tungsten oxides within a Teflon chamber were used to detect methyl propyl sulfide and 2-nonanone emitted from rancid onions up to 196 and 145 ppm, respectively [96]. TiO_2, NiO, and CeO_2 nanoparticles have been used in the fabrication of an electronic tongue to oversee the phenolic maturity of grapes and evaluate their palatability. These MOx Nanoparticle Sensors (MONS) can systemically respond to phenolic acids and polyphenols present in grape wines in concentrations as low as 10 nM [97].

Alcohol sensors have an impact in food packaging industries, fermentation process, wine preparation, and medical applications. It is important that the MOx sensor uses a lower operating temperature for detection of the alcohol due to the volatility of the analyte. Several MOx have been investigated for detection of alcohol. A sensing strategy involving MOx-based molecularly imprinted polymer electrodes was employed for quantification of alcohol.

The sensor used Ag-doped $LaFeO_3$ (SLMIPs) prepared via sol-gel method employing molecularly imprinted polymers as precursors. Selective detection of methanol vapors with a detection limit of 5 ppm was achieved from the sensor [98].

6.6.5 MOx Sensors in Agriculture

Chemical sensors based on MOx have been employed to detect a wide range of pesticides. TiO_2 nanoparticles modified on screen-printed carbon electrodes (SPE) have been utilized for sensing of dichlofenthion with a detection limit of 2 nM. Green vegetable samples have been tested with the modified electrode with good recovery values and in good agreement with the gas chromatography–mass spectrometry analysis [99]. SnO_2 doped with Cr and nickel was prepared and used as a specific matrix for detecting chlorpyrifos with an observed linear range of 0.01–100 ppm and a detection limit of 10 ppb [100]. A combination of ZrO_2 nanoparticles synthesized electrodynamically on an Au electrode was used for the detection of various nitrocompounds like methyl parathion, fenitrothion, and paraoxon. The sensor exhibited a linear range of 5–100 ng ml^{-1} and a detection limit of the prepared sensor was 3 ng ml^{-1}. The adsorption time for each addition of pesticide was maintained as 10 minutes [101].

6.6.6 MOx Sensors for Hazard Analysis

The gases emitted due to deployment of CWAs are a mixture of hydrocarbons and phosphonates that can be detected by MOx-based devices. Semiconducting MOx arrays (SMOs) comprising indium, tin, tungsten, yttrium, and copper oxide thin films collectively deposited through a drop-coating method has facilitated the distinguishing of various constituents of combustion with enhanced sensitivity. Concentrations as low as 10 ppb were easily quantified by this system [102]. Another need emerging due to terrorist activities in the modern era is the development of a highly sensitive and portable sensing strategy for detection of concealed explosives. The challenge in explosive detection is the need for rapid and accurate detection in the vapor phase owing to their low vapor pressure [103]. Indium-doped zinc oxide was employed to detect trinitrotoluene (TNT), dinitrotoluene (DNT), picric acid (PA), RDX, and paranitrotoluene (PNT) [104]. The specificity was regulated by altering the degree of doping. The sensor array worked at room temperature and exhibited a rapid response below 6.3 seconds and recovery time below 14 seconds. However, the presence of moisture interfered with the sensor performance thereby limiting the utility of this system. The current research in this field is focused on integrating different sensing elements to form an array for rapid detection of trace quantities of different explosives.

A comparison of the response of various sensors toward identification of fire arising from different conditions (open or smouldering) and sources (wood, polymer, cotton, etc.) revealed that SnO_2 or WO_3 sensors exhibited the fastest response for most types of fires [105]. The study also revealed that the location size of the fire, oxygen content, and temperatures reached will also influence the performance of the sensors. Another study compared the sensitivity of an array comprising eight MOx sensors and three electrochemical sensors to detect the smoke components arising from burning electrical insulation [106]. The MOx sensors were able to detect the insulation combustion products and principal component analysis was performed to identify the type of insulation. In a recent attempt, MOx gas sensors were trained to detect low concentrations of the plant-based organic molecules namely, eucalyptol, furfural, α-pinene, and 2-methoxyphenol, that are released from a burning eucalyptus tree *Eucalyptus globus*. This strategy will be very useful for early detection of forest fires and prevent loss of forest cover [107]. It is evident that there exists much scope for development of a portable and reliable detection unit to detect fire hazards. But, several in-depth studies are required to understand the influence of environmental parameters and interferents on the sensor performance.

6.6.7 Flexible Sensors Based on MOx

Flexible sensors are of topical interest as they are lightweight, can withstand shock, and are more amenable to surface modifications thereby having applications transcending different domains. Several flexible sensor devices comprising MOx as a key sensing element have been reported in recent years. A triple component biosensor comprising high density platinum nanoparticles ultrasonically electrodeposited on reduced graphene oxide (rGO) paper carrying MnO_2 nanowires was reported for the nonenzymatic detection of H_2O_2 in live cells. Apart from the enhanced electrocatalytic behavior of the hybrid network, the absence of size limits of MnO_2, Pt, and rGO improves the scalability of the sensor in practical applications. The sensor demonstrated a sensitivity of $129.5\,\mu Acm^{-2}\,mM^{-1}$, LoD of $1\,\mu M$, and a response time of three seconds. Upon inducing mechanical stress by bending (for 100 times), the detection of H_2O_2 altered by less than 5% implying the negligible impact of bending and mechanical stress on the sensor performance suggesting its applicability in the development of microdiagnostic devices [108]. The field of flexible sensors is slowly developing and further advancements in this niche area are expected in the coming years.

6.6.8 MOx-Based Lab-on-a-Chip Sensors

Miniaturized microfluidic sensors have elicited much interest in the twenty-first century due to the possibilities of a myriad of applications arising due to their small size and portability. These chips can be easily integrated in devices thereby leading to development of mobile sensors that can revolutionalize point-of-care diagnosis of patients and environmental monitoring [12]. A microfluidic device for environmental monitoring reported in recent literature employed an integrated screen printed electrode drop-casted with zinc and bismuth oxide. The device showed a good linear relation toward detection of the metal ion pollutants with a recovery time of 100–150 seconds [109]. Recently, 3D printing has been used for the fabrication of variety of moulds for fabrication of sensor and microfluidic devices. VOCs were monitored using a 3D printed artificial nose, which was found to be hundred times more sensitive than the human nose and could sense the toxic gases present in the environment within seconds [110]. This emerging field promises to accelerate the development of MOx-based sensing devices by 3D printing and may well represent the future of the next generation of "smart sensors" in the immediate future.

6.7 Concluding Remarks

MOx represent a wide range of materials that respond to changes in the adsorbed layer on their surface with very high sensitivity. The ease of synthesis of these oxides and the ability to manipulate their morphology at the nanoscale has resulted in significant improvement in their sensing characteristics as well as a reduction in their operating temperatures. Though most MOx-based chemical sensors have been extensively used for detection of target molecules in the gas phase, doping and surface modification strategies has expanded the repertoire of MOx-based sensors to detection in the liquid phase as well. The challenges of specificity of detection are being addressed with appropriate doping strategies and surface functionalization approaches. The portability and stability of MOx sensors are key contributors for their integration in new age technologies such as robots, smart devices, miniaturized sensor chips, and flexible and wearable sensors that are expected to be commercialized in the near future. The current trend in sensing is to detect multiple analytes simultaneously which can be easily achieved by designing MOx-based sensor arrays. It is evident that with the ever-increasing demand for affordable real-time monitoring devices, MOx-based sensors are destined to make a mark in all industrial, commercial, and domestic segments.

Acknowledgment

The authors thank RCUK (MRC/P027881) and SASTRA Deemed University for funding and infrastructural support. UMK also acknowledges University of Bordeaux for the visiting professorship. The authors also thank Ms. Kaviarasi VS for helping with the artwork.

References

1 Dahman, Y. (2017). *Nanotechnology and Functional Materials for Engineers*, 1–268. Elsevier.

2 Sun, Y.-F., Liu, S.-B., Meng, F.-L. et al. (2012). Metal oxide nanostructures and their gas sensing properties: a review. *Sensors (Basel, Switzerland)* 12 (3): 2610–2631.

3 Fernandez, A.C., Sakthivel, P., and Jesudurai, J. (2018). Semiconducting metal oxides for gas sensor applications. *Journal of Materials Science: Materials in Electronics* 29 (1): 357–364.

4 Berna, A. (2010). Metal oxide sensors for electronic noses and their application to food analysis. *Sensors (Basel, Switzerland)* 10 (4): 3882–3910.

5 Wang, C., Yin, L., Zhang, L. et al. (2010). Metal oxide gas sensors: sensitivity and influencing factors. *Sensors* 10 (3): 2088–2106.

6 Kanan, M.S., El-Kadri, M.O., Abu-Yousef, A.I. et al. (2009). Semiconducting metal oxide based sensors for selective gas pollutant detection. *Sensors* 9 (10): 8158–8196.

7 Ponzoni, A., Baratto, C., Cattabiani, N. et al. (2017). Metal oxide gas sensors, a survey of selectivity issues addressed at the SENSOR lab, Brescia (Italy). *Sensors* 17 (4): 714–742.

8 Praveen, K., Sekhar, E.L.B., Mukundan, R., and Garzon, F.H. (2010). Chemical sensors for environmental monitoring and homeland security. *The Electrochemical Society Interface* 19 (4): 35–40.

9 Yu, H.-D., Regulacio, M.D., Ye, E. et al. (2013). Chemical routes to top-down nanofabrication. *Chemical Society Reviews* 42 (14): 6006–6018.

10 Zhang, B. and Gao, P.-X. (2019). Metal oxide nanoarrays for chemical sensing: a review of fabrication methods, sensing modes, and their inter-correlations. *Frontiers in Materials* 6: 55.

11 Francioso, L. (2014). 5 – Chemiresistor gas sensors using semiconductor metal oxides. In: *Nanosensors for Chemical and Biological Applications* (ed. K.C. Honeychurch), 101–124. Woodhead Publishing.

12 Yoon, J.-W. and Lee, J.-H. (2017). Toward breath analysis on a chip for disease diagnosis using semiconductor-based chemiresistors: recent progress and future perspectives. *Lab on a Chip* 17 (21): 3537–3557.

13 Xu, J., Pan, Q., Shun, Y.A. et al. (2000). Grain size control and gas sensing properties of ZnO gas sensor. *Sensors and Actuators B: Chemical* 66 (1): 277–279.

14 Lin, R.-B., Liu, S.-Y., Ye, J.-W. et al. (2016). Photoluminescent metal–organic frameworks for gas sensing. *Advanced Science* 3 (7): 1500434. (pp. 1–20).

15 Basu, S. and Basu, P.K. (2018). Nanocrystalline metal oxides for methane sensors: role of noble metals. *Journal of Sensors* 2009: 1–20.

16 Lin, T., Lv, X., Hu, Z. et al. (2019). Semiconductor metal oxides as chemoresistive sensors for detecting volatile organic compounds. *Sensors* 19 (2): 233–264.

17 Shruthi, J., Jayababu, N., and Reddy, M.V.R. (2019). Room temperature ethanol gas sensing performance of $CeO_2-In_2O_3$ heterostructured nanocomposites. *AIP Conference Proceedings* 2082 (1): 030020.

18 Luo, Y., Zhang, C., Zheng, B. et al. (2017). Hydrogen sensors based on noble metal doped metal-oxide semiconductor: a review. *International Journal of Hydrogen Energy* 42 (31): 20386–20397.

19 Li, S., Zhang, H., Xu, J. et al. (2005). Hydrothermal synthesis of flower-like $SrCO_3$ nanostructures. *Materials Letters* 59: 420–422.

20 Vander Wal, R.L., Hunter, G.W., Xu, J.C. et al. (2009). Metal-oxide nanostructure and gas-sensing performance. *Sensors and Actuators B: Chemical* 138 (1): 113–119.

21 D'Amico, A., Ferri, G., and Zompanti, A. (2019). Sensor systems for breathprinting: a review of the current technologies for exhaled breath analysis based on a sensor array with the aim of integrating them in a standard and shared procedure. In: *Breath Analysis* (eds. G. Pennazza and M. Santonico), 49–79.

22 Usubharatana, P., McMartin, D., Veawab, A. et al. (2006). Photocatalytic process for CO_2 emission reduction from industrial flue gas streams. *Industrial & Engineering Chemistry Research* 45 (8): 2558–2568.

23 Traversa, E. (1995). Ceramic sensors for humidity detection: the state-of-the-art and future developments. *Sensors and Actuators B: Chemical* 23: 135–156.

24 Hellström, M., Quaranta, V., and Behler, J. (2019). One-dimensional vs. two-dimensional proton transport processes at solid–liquid zinc-oxide–water interfaces. *Chemical Science* 10 (4): 1232–1243.

25 Zhang, D., Chang, H., Li, P. et al. (2015). Fabrication and characterization of an ultrasensitive humidity sensor based on metal oxide/graphene hybrid nanocomposite. *Sensors and Actuators B: Chemical* 225: 233–240.

26 Li, Z., Li, H., Wu, Z. et al. (2019). Advances in designs and mechanisms of semiconducting metal oxide nanostructures for high-precision gas sensors operated at room temperature. *Materials Horizons* 6 (3): 470–506.

27 Binions, R. and Naik, A. (2013). Metal oxide semiconductor gas sensors in environmental monitoring. In: *Semiconductor Gas Sensors*, 433–466. Woodhead Publishing.

28 Barsan, N., Koziej, D., and Weimar, U. (2007). Metal oxide-based gas sensor research. *Sensors and Actuators B: Chemical* 121: 18–35.

29 Akamatsu, T., Itoh, T., Izu, N. et al. (2013). NO and NO_2 sensing properties of WO_3 and Co_3O_4 based gas sensors. *Sensors* 13 (9): 12467–12481.

30 Karunagaran, B., Uthirakumar, P., Chung, S.J. et al. (2007). TiO_2 thin film gas sensor for monitoring ammonia. *Materials Characterization* 58: 680–684.

31 Zhang, Z., Haq, M., Wen, Z. et al. (2018). Ultrasensitive ppb-level NO_2 gas sensor based on WO_3 hollow nanosphers doped with Fe. *Applied Surface Science* 434: 891–897.

32 Jung, S., Baik, K.H., Ren, F. et al. (2018). AlGaN/GaN heterostructure based schottky diode sensors with ZnO nanorods for environmental ammonia monitoring applications. *ECS Journal of Solid State Science and Technology* 7 (7): 3020–3024.

33 Liao, L., Zhang, Z., Yan, B. et al. (2009). Multifunctional CuO nanowire devices: p-type field effect transistors and CO gas sensors. *Nanotechnology* 20 (8): 085203.

34 Hjiri, M., El Mir, L., Leonardi, S.G. et al. (2014). Al-doped ZnO for highly sensitive CO gas sensors. *Sensors and Actuators B: Chemical* 196: 413–420.

35 Chang, S.-J., Hsueh, T.-J., Chen, I.C. et al. (2008). Highly sensitive ZnO nanowire CO sensors with the adsorption of Au nanoparticles. *Nanotechnology* 19 (17): 175502.

36 Neri, G., Leonardi, S.G., Latino, M. et al. (2013). Sensing behavior of SnO_2/ reduced graphene oxide nanocomposites toward NO_2. *Sensors and Actuators B: Chemical* 179: 61–68.

37 Shinde, S.D., Patil, G.E., Kajale, D.D. et al. (2012). Synthesis of ZnO nanorods by spray pyrolysis for H_2S gas sensor. *Journal of Alloys and Compounds* 528: 109–114.

38 Mani, G.K. and Rayappan, J.B.B. (2013). A highly selective room temperature ammonia sensor using spray deposited zinc oxide thin film. *Sensors and Actuators B: Chemical* 183: 459–466.

39 Ganesh, R.S., Durgadevi, E., Navaneethan, M. et al. (2018). Tuning the selectivity of NH_3 gas sensing response using Cu-doped ZnO nanostructures. *Sensors and Actuators A: Physical* 269: 331–341.

40 Modafferi, V., Panzera, G., Donato, A. et al. (2012). Highly sensitive ammonia resistive sensor based on electrospun V_2O_5 fibers. *Sensors and Actuators B: Chemical* 163 (1): 61–68.

41 Feng, Q., Li, X., Wang, J. et al. (2016). Reduced graphene oxide (rGO) encapsulated Co_3O_4 composite nanofibers for highly selective ammonia sensors. *Sensors and Actuators B: Chemical* 222: 864–870.

42 Dar, G.N., Umar, A., Zaidi, S.A. et al. (2012). Ultra-high sensitive ammonia chemical sensor based on ZnO nanopencils. *Talanta* 89: 155–161.

43 Zhang, J., Wang, S., Xu, M. et al. (2009). Polypyrrole-coated SnO_2 hollow spheres and their application for ammonia sensor. *The Journal of Physical Chemistry C* 113 (5): 1662–1665.

44 Zhang, D., Jiang, C., Liu, J. et al. (2017). Carbon monoxide gas sensing at room temperature using copper oxide-decorated graphene hybrid nanocomposite prepared by layer-by-layer self-assembly. *Sensors and Actuators B: Chemical* 247: 875–882.

45 Wang, X., Qin, H., Sun, L. et al. (2013). CO_2 sensing properties and mechanism of nanocrystalline $LaFeO_3$ sensor. *Sensors and Actuators B: Chemical* 188: 965–971.

46 Alharbi, N.D., Ansari, M.S., Salah, N. et al. (2016). Zinc oxide-multi walled carbon nanotubes nanocomposites for carbon monoxide gas sensor application. *Journal of Nanoscience and Nanotechnology* 16 (1): 439–447.

47 Su, P.-G. and Peng, S.-L. (2015). Fabrication and NO_2 gas-sensing properties of reduced graphene oxide/WO_3 nanocomposite films. *Talanta* 132: 398–405.

48 Navale, S.T., Bandgar, D.K., Nalage, S.R. et al. (2013). Synthesis of Fe_2O_3 nanoparticles for nitrogen dioxide gas sensing applications. *Ceramics International* 39 (6): 6453–6460.

49 Mane, A.T., Navale, S.T., Sen, S. et al. (2015). Nitrogen dioxide (NO_2) sensing performance of p-polypyrrole/n-tungsten oxide hybrid nanocomposites at room temperature. *Organic Electronics* 16: 195–204.

50 Navale, Y.H., Navale, S.T., Ramgir, N.S. et al. (2017). Zinc oxide hierarchical nanostructures as potential NO_2 sensors. *Sensors and Actuators B: Chemical* 251: 551–563.

51 Das, S., Rana, S., Mursalin, S.M. et al. (2015). Sonochemically prepared nanosized $BiFeO_3$ as novel SO_2 sensor. *Sensors and Actuators B: Chemical* 218: 122–127.

52 Tyagi, P., Sharma, A., Tomar, M. et al. (2017). A comparative study of $RGO-SnO_2$ and $MWCNT-SnO_2$ nanocomposites based SO_2 gas sensors. *Sensors and Actuators B: Chemical* 248: 980–986.

53 Marikutsa, A., Rumyantseva, M., Baranchikov, A. et al. (2015). Nanocrystalline $BaSnO_3$ as an alternative gas sensor material: surface reactivity and high sensitivity to SO_2. *Materials* 8 (9): 6437–6454.

54 Palimar, S., Kaushik, S.D., Siruguri, V. et al. (2016). Investigation of Ca substitution on the gas sensing potential of $LaFeO_3$ nanoparticles towards low concentration SO_2 gas. *Dalton Transactions* 45 (34): 13547–13555.

55 Shendage, S.S., Patil, V.L., Vanalakar, S.A. et al. (2017). Sensitive and selective NO_2 gas sensor based on WO_3 nanoplates. *Sensors and Actuators B: Chemical* 240: 426–433.

56 Wang, Y., Cui, X., Yang, Q. et al. (2016). Preparation of Ag-loaded mesoporous WO_3 and its enhanced NO_2 sensing performance. *Sensors and Actuators B: Chemical* 225: 544–552.

57 Kneer, J., Knobelspies, S., Bierer, B. et al. (2016). New method to selectively determine hydrogen sulfide concentrations using CuO layers. *Sensors and Actuators B: Chemical* 222: 625–631.

58 Xiao, T., Wang, X.-Y., Zhao, Z.-H. et al. (2014). Highly sensitive and selective acetone sensor based on C-doped WO_3 for potential diagnosis of diabetes mellitus. *Sensors and Actuators B: Chemical* 199: 210–219.

59 Kim, S.-J., Choi, S.-J., Yang, D.-J. et al. (2014). Highly sensitive and selective hydrogen sulfide and toluene sensors using Pd functionalized WO_3 nanofibers for potential diagnosis of halitosis and lung cancer. *Sensors and Actuators B: Chemical* 193: 574–581.

60 Moon, H.G., Jung, Y., Han, S.D. et al. (2018). All villi-like metal oxide nanostructures-based chemiresistive electronic nose for an exhaled breath analyzer. *Sensors and Actuators B: Chemical* 257: 295–302.

61 Obisesan, O.R., Adekunle, A.S., Oyekunle, J.A.O. et al. (2019). Development of electrochemical nanosensor for the detection of malaria parasite in clinical samples. *Frontiers in Chemistry* 7 (89): 1–15.

62 Qi, Q., Zhang, T., Liu, L. et al. (2008). Selective acetone sensor based on dumbbell-like ZnO with rapid response and recovery. *Sensors and Actuators B: Chemical* 134 (1): 166–170.

63 Lakhane, M., Khairnar, R., and Mahabole, M. (2016). Metal oxide blended ZSM-5 nanocomposites as ethanol sensors. *Bulletin of Materials Science* 39 (6): 1483–1492.

64 Gu, C., Guan, W., Liu, X. et al. (2017). Controlled synthesis of porous Ni-doped SnO$_2$ microstructures and their enhanced gas sensing properties. *Journal of Alloys and Compounds* 692: 855–864.

65 Vaishnav, V.S., Patel, S.G., and Panchal, J.N. (2015). Development of ITO thin film sensor for detection of benzene. *Sensors and Actuators B: Chemical* 206: 381–388.

66 Rahman, M.M., Alam, M.M., and Asiri, A.M. (2018). Carbon black co-adsorbed ZnO nanocomposites for selective benzaldehyde sensor development by electrochemical approach for environmental safety. *Journal of Industrial and Engineering Chemistry* 65: 300–308.

67 Tang, W. and Wang, J. (2015). Mechanism for toluene detection of flower-like ZnO sensors prepared by hydrothermal approach: charge transfer. *Sensors and Actuators B: Chemical* 207: 66–73.

68 Fang, C., Wang, S., Wang, Q. et al. (2010). Coralloid SnO$_2$ with hierarchical structure and their application as recoverable gas sensors for the detection of benzaldehyde/acetone. *Material Chemistry and Physics* 122: 30–34.

69 Sun, Y., Wei, Z., Zhang, W. et al. (2016). Synthesis of brush-like ZnO nanowires and their enhanced gas-sensing properties. *Journal of Materials Science* 51 (3): 1428–1436.

70 Gao, H., Wei, D., Lin, P. et al. (2017). The design of excellent xylene gas sensor using Sn-doped NiO hierarchical nanostructure. *Sensors and Actuators B: Chemical* 253: 1152–1162.

71 Hubalek, J., Malysz, K., Prášek, J. et al. (2004). Pt-loaded Al$_2$O$_3$ catalytic filters for screen-printed WO$_3$ sensors highly selective to benzene. *Sensors and Actuators B: Chemical* 101: 277–283.

72 Woo, H.-S., Kwak, C.-H., Chung, J.-H. et al. (2015). Highly selective and sensitive xylene sensors using Ni-doped branched ZnO nanowire networks. *Sensors and Actuators B: Chemical* 216: 358–366.

73 Akiyama, T., Ishikawa, Y., and Hara, K. (2013). Xylene sensor using double-layered thin film and Ni-deposited porous alumina. *Sensors and Actuators B: Chemical* 181: 348–352.

74 Li, F., Guo, S., Shen, J. et al. (2017). Xylene gas sensor based on Au-loaded WO$_3$·H$_2$O nanocubes with enhanced sensing performance. *Sensors and Actuators B: Chemical* 238: 364–373.

75 Güntner, A.T., Pineau, N.J., Chie, D. et al. (2016). Selective sensing of isoprene by Ti-doped ZnO for breath diagnostics. *Journal of Materials Chemistry B* 4 (32): 5358–5366.

76 Park, Y., Yoo, R., Park, S.R. et al. (2019). Highly sensitive and selective isoprene sensing performance of ZnO quantum dots for a breath analyzer. *Sensors and Actuators B: Chemical* 290: 258–266.

77 Herrera-Rivera, R., Olvera, M.d.l.L., and Maldonado, A. (2017). Propane sensor pellets based on nanopowders Cu-doped ZnO. In: *2017 14th International Conference on Electrical Engineering, Computing Science and Automatic Control (CCE) (20–22 October 2017)*, 1–5. IEEE.

78 Han, B., Liu, X., Xing, X. et al. (2016). A high response butanol gas sensor based on ZnO hollow spheres. *Sensors and Actuators B: Chemical* 237: 423–430.

79 Malik, R., Tomer, V.K., Chaudhary, V. et al. (2017). Ordered mesoporous In-(TiO$_2$/WO$_3$) nanohybrid: an ultrasensitive *n*-butanol sensor. *Sensors and Actuators B: Chemical* 239: 364–373.

80 Yang, B., Liu, J., Qin, H. et al. (2018). PtO$_2$-nanoparticles functionalized CuO polyhedrons for *n*-butanol gas sensor application. *Ceramics International* 44 (9): 10426–10432.

81 Wang, Y., Zhang, B., Liu, J. et al. (2016). Au-loaded mesoporous WO$_3$: preparation and *n*-butanol sensing performances. *Sensors and Actuators B: Chemical* 236: 67–76.

82 Sohn, J.R., Park, H.D., and Lee, D.D. (2000). Infrared spectroscopic study of acetonitrile on SnO$_2$-based thick film and its characteristics as a gas sensor. *Journal of Catalysis* 195 (1): 12–19.

83 Guo, D., Cai, P., Sun, J. et al. (2016). Reduced-graphene-oxide/metal-oxide p-n heterojunction aerogels as efficient 3D sensing frameworks for phenol detection. *Carbon* 99: 571–578.

84 Rahman, M.M., Balkhoyor, H.B., and Asiri, A.M. (2017). Phenolic sensor development based on chromium oxide-decorated carbon nanotubes for environmental safety. *Journal of Environmental Management* 188: 228–237.

85 Xu, X., Arab Pour Yazdi, M., Rauch, J.-Y. et al. (2015). Tungsten oxide thin films sputter deposited by the reactive gas pulsing process for the dodecane detection. *Materials Today: Proceedings* 2 (9, Part B): 4656–4663.

86 Xu, X., Arab Pour Yazdi, M., Sanchez, J.-B. et al. (2018). Exploiting the dodecane and ozone sensing capabilities of nanostructured tungsten oxide films. *Sensors and Actuators B: Chemical* 266: 773–783.

87 Zhang, Y., Li, D., Qin, L. et al. (2018). Preparation of Au-loaded TiO$_2$ pecan-kernel-like and its enhanced toluene sensing performance. *Sensors and Actuators B: Chemical* 255: 2240–2247.

88 Gürbüz, E. and Şahin, B. (2018). Zn-doping to improve the hydration level sensing performance of CuO films. *Applied Physics A* 124: 795. (pp. 1–9).

89 Fernandes, D.M., Costa, M., Pereira, C. et al. (2014). Novel electrochemical sensor based on *N*-doped carbon nanotubes and Fe$_3$O$_4$ nanoparticles: simultaneous voltammetric determination of ascorbic acid, dopamine and uric acid. *Journal of Colloid and Interface Science* 432: 207–213.

90 Khan, R., Kaushik, A., Solanki, P.R. et al. (2008). Zinc oxide nanoparticles-chitosan composite film for cholesterol biosensor. *Analytica Chimica Acta* 616 (2): 207–213.

91 Bano, K., Bajwa, S., Bassous, N.J. et al. (2019). Development of biocompatible 1D CuO nanoneedles and their potential for sensitive, mass-based detection of anti-tuberculosis drugs. *Applied Nanoscience* 9 (6): 1341–1351.

92 Alavi-Tabari, S.A.R., Khalilzadeh, M.A., and Karimi-maleh, H. (2018). Simultaneous determination of doxorubicin and dasatinib as two breast anticancer drugs uses an amplified sensor with ionic liquid and ZnO nanoparticle. *Journal of Electroanalytical Chemistry* 811: 84–88.

93 Ansari, S., Ansari, M.S., Satsangee, S.P. et al. (1046). WO_3 decorated graphene nanocomposite based electrochemical sensor: a prospect for the detection of anti-anginal drug. *Analytica Chimica Acta* 2019: 99–109.

94 Gupta, V.K., Karimi-Maleh, H., Agarwal, S. et al. (2018). Fabrication of a food nano-platform sensor for determination of vanillin in food samples. *Sensors* 18 (9): 2817–2825.

95 Gobbi, E., Falasconi, M., Zambotti, G. et al. (2015). Rapid diagnosis of Enterobacteriaceae in vegetable soups by a metal oxide sensor based electronic nose. *Sensors and Actuators B: Chemical* 207: 1104–1113.

96 Konduru, T., Rains, G.C., and Li, C. (2015). A customized metal oxide semiconductor-based gas sensor array for onion quality evaluation: system development and characterization. *Sensors* 15 (1): 1252–1273.

97 Garcia-Hernandez, C., Medina-Plaza, C., Garcia-Cabezon, C. et al. (2018). Monitoring the phenolic ripening of red grapes using a multisensor system based on metal-oxide nanoparticles. *Frontiers of Chemistry* 6: 131–138.

98 Rong, Q., Zhang, Y., Lv, T. et al. (2018). Highly selective and sensitive methanol gas sensor based on molecular imprinted silver-doped $LaFeO_3$ core–shell and cage structures. *Nanotechnology* 29 (14): 145503.

99 Li, H., Li, J., Yang, Z. et al. (2011). A novel photoelectrochemical sensor for the organophosphorus pesticide dichlofenthion based on nanometer-sized titania coupled with a screen-printed electrode. *Analytical Chemistry* 83 (13): 5290–5295.

100 Khan, N., Athar, T., Fouad, H. et al. (2017). Application of pristine and doped SnO_2 nanoparticles as a matrix for agro-hazardous material (organophosphate) detection. *Scientific Reports* 7: 42510.

101 Liu, G. and Lin, Y. (2005). Electrochemical sensor for organophosphate pesticides and nerve agents using zirconia nanoparticles as selective sorbents. *Analytical Chemistry* 77 (18): 5894–5901.

102 Tomchenko, A., Harmer, G.P., and Marquis, B.T. (2005). Detection of chemical warfare agents using nanostructured metal oxide sensors. *Sensors and Actuators B: Chemical* 108: 41–55.

103 Senesac, L. and Thundat, T.G. (2008). Nanosensors for trace explosive detection. *Materials Today* 11 (3): 28–36.

104 Ge, Y., Wei, Z., Li, Y. et al. (2017). Highly sensitive and rapid chemiresistive sensor towards trace nitro-explosive vapors based on oxygen vacancy-rich and defective crystallized In-doped ZnO. *Sensors and Actuators B: Chemical* 244: 983–991.

105 Baccar, H., Thamri, A., Clément, P. et al. (2015). Pt- and Pd-decorated MWCNTs for vapour and gas detection at room temperature. *Beilstein Journal of Nanotechnology* 6 (1): 919–927.

106 Ni, M., Stetter, J.R., and Buttner, W.J. (2008). Orthogonal gas sensor arrays with intelligent algorithms for early warning of electrical fires. *Sensors and Actuators B: Chemical* 130: 889–899.

107 Paczkowski, S., Nicke, S., Ziegenhagen, H. et al. (2015). Volatile emission of decomposing pig carcasses (*Sus scrofa domesticus L.*) as an indicator for the postmortem interval. *Journal of Forensic Sciences* 60 (s1): 130–137.

108 Xiao, F., Li, Y., Zan, X. et al. (2012). Growth of metal–metal oxide nanostructures on freestanding graphene paper for flexible biosensors. *Advanced Functional Materials* 22 (12): 2487–2494.

109 Frau, I., Wylie, S., Cullen, J. et al. (2019). Microwaves and functional materials: a novel method to continuously detect metal ions in water. In: *Modern Sensing Technologies* (eds. S.C. Mukhopadhyay, K.P. Jayasundera and O.A. Postolache), 179–201. Cham: Springer International Publishing.

110 Wilson, A.D. (2018). Application of electronic-nose technologies and VOC-biomarkers for the noninvasive early diagnosis of gastrointestinal diseases (†). *Sensors (Basel, Switzerland)* 18 (8): 2613.

7

Metal Oxide Gas Sensor Electronic Interfaces

Zeinab Hijazi[1,2], Daniele D. Caviglia[1], and Maurizio Valle[1]

[1] COSMIC Lab, University of Genova, Genova, Italy
[2] Department of Electric and Electronic Engineering, International University of Beirut (BIU), Beirut, Lebanon

7.1 General Introduction

7.1.1 Gas Sensing System

Through the nineteenth and early twentieth centuries, canary bird was used as an early detection system against life-threatening gases. Coal miners used to bring canaries down to the mines with them. The canary is a very songful bird; once it stopped singing or even dies, the miners receive a signal about the existence of toxic gases such as carbon dioxide, carbon monoxide, and methane, indicating to exit the mine quickly [1]. Because of the modern safety regulations and tightening emission, detecting explosive gases (such as hydrogen), as well as, toxic and/or harmful gases (such as Nitrogen Oxide and Carbon monoxide) is attracting a large research field.

Electronic nose (e-nose) or gas sensing system has attracted many recent research due to its importance in many application domains, such as industrial applications [2], monitoring indoor air quality [3], medical care [4–6], custom security [7], controlling food quality [8–14], environmental air quality monitoring [15–18], military [19], in addition to dangerous gas detection [20, 21]. New opportunities for sensing devices involve hot topics such as internet of things (IoT) devices and mobile platforms, i.e. wearable and smart phones. These applications require a portable e-nose system where its main metrics are: low cost, low power, small size, long time smelling ability, as well as the capability of dangerous gas exposure.

Smart Sensors for Environmental and Medical Applications, First Edition. Edited by Hamida Hallil and Hadi Heidari.
© 2020 The Institute of Electrical and Electronics Engineers, Inc.
Published 2020 by John Wiley & Sons, Inc.

Gas sensors measure the concentration of gas in the ambient atmosphere. To best understand the gas sensing process, the gas sensing system is considered as an artificial olfactory system also known as an electronic nose or e-nose [22, 23]. Mimicking the mammalian olfactory system, olfactory cells, olfactory neurons, and cerebrum stands for gas sensor array, signal conditioning circuit, and pattern recognition system in e-nose systems respectively. Towards implementing a portable and low power hand-held gas sensing system, this chapter targets the design of integrated electronic signal conditioning circuits also known as integrated electronic readout circuits. In this senario, this chapter, reports first a study of the available gas sensing technologies focusing on the advantages and limitations of each, then metal oxide (MOX)-based gas sensors will be introduced as a sensing technology for portable handheld gas detection used for environmental monitoring. Next, the principle of operation of such sensors will be addressed. Characterization will be then provided to define the specifications of the electronic integrated readout circuit. Thereafter, a state-of-the-art study will be provided to introduce the MOX gas sensor electronic interface circuits available in literature. The requirements of each block at transistor level design will be discussed later. High power consumption represents a main barrier for MOX based gas sensory systems. Efforts done in terms of power consumption reduction for portable mobile applications, will be addressed. Conclusions and future perspectives will be finally reported.

7.1.2 Gas Sensing Technologies

Different studies established various branches of gas sensing technologies. Chemical sensing technologies are those exploiting chemical interactions to detect a targeted substance. Each technology suits a specific application domain. Various types of gas sensors are available on the market nowadays. Table 7.1 compares several types of gas sensors highlighting their advantages and limitations [20]. Low cost, long life time, high sensitivity, small size, in addition to simplicity of operation, flexibility in production, and compatibility with the standard CMOS process put chemoresistive semiconductor gas sensors based on MOX in a preceding place compared to electrochemical, carbon nanotubes, optical, and other types of gas sensors. MOX-based gas sensors represent a candidate for realizing integrated and miniaturized chemoresistive thin film sensors with long-term stability without maintenance. However, MOX based sensors are poorly selective; such sensors respond in a similar way to different oxidizing and reducing gases. To improve their selectivity, semiconductor sensors based on MOX can be arranged as large matrix arrays of similar sensing materials [18] or as heterogeneous microsensor array of different sensing materials in discrete or integrated form [24].

Table 7.1 Basic gas sensors types: advantages and limitations.

Gas sensor type	Advantages	Limitations
Chemoresistive	Low cost Long life time Small size High sensitivity *Wide range of gases* Short response time	*High energy consumption* Poor selectivity
Electrochemical (polymer)	low cost *High sensitivity* *Short response time* *Portable and simple structure*	*Irreversibility* *Long life time instability* *Poor selectivity*
Carbon nanotubes	*Small size* *Ultra-sensitive* *Short response time* *Great adsorptive capacity* *Large surface area to volume ratio*	*High cost* *Difficulties in fabrication and repeatability*
Chemo-optical	High sensitivity Excellent selectivity Long life time Insensitive to environmental changes	*High cost* *Large size* *Difficulty in miniaturization*
Calorimetric methods	*Low cost* *Adequate sensitivity for industrial detection* *Stable at ambient temperature*	*Intrinsic deficiencies in selectivity* *Risk of catalyst poisoning and explosion*
Chromatograph	*High selectivity* *High sensitivity* *Excellent separation performance*	High cost Difficulty in miniaturization for portable applications

7.2 MOX Gas Sensors

7.2.1 Principle of Operation

MOX resistive gas sensors are also known as chemoresistive sensors [25]. In the presence of a gas and at elevated temperature ranging between 150 and 400 °C, chemical adsorption of oxygen species in air at the surface of the semiconductor material takes place, creating surface acceptor states (O^-, O^{2-}, O_2^-). Electrons are

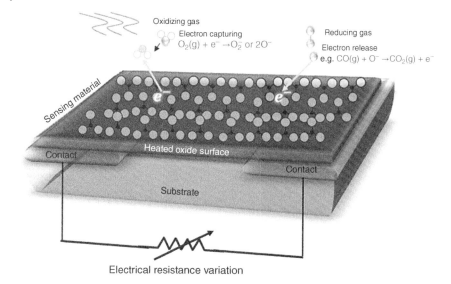

Figure 7.1 Gas adsorption mechanism at the surface of heated MOX sensing films.

then trapped near the surface, thus, the electron depletion region increases. Thereby, oxidation/reduction reaction starts at the surface of the sensor's resistive thin films. Therefore, the electrical conductivity of the oxide surface is modified, and, the transducer of the sensor transforms the gas concentration variation into an electrical resistance variation. Figure 7.1 shows the gas adsorption mechanism at the surface of heated n-type oxide causing an electric resistance variation.

Common materials for chemoresistive MOX sensors are SnO_2, TiO_2, WO_3, and ZnO [24]. Both n-type (such as WO_3, SnO_2, ZnO, TiO_2) and p-type (such as NiO, CuO) MOX semiconductors can be used as sensing materials. Note that the n-type oxide semiconductors are used up to 90% compared to p-types oxides [26], since the mobility of charge carriers for n-type are higher than that of the p-type.

7.2.2 Assessment of Available MOX-Based Gas Sensors

It is important to characterize some MOX-based gas sensors to better understand the factors affecting their performance and to set the main requirements of the electronic readout circuit. Table 7.2 provides the main characteristics of some commercial and noncommercial (prototype) MOX gas sensors. Commercial gas sensors are tailored to particular gases and concentration levels specified for a defined application, whereas noncommercial gas sensors are laboratory-developed

Table 7.2 MOX based gas sensors characterization.

Sensors type	Commercial						Noncommercial (prototype)		
	MSGS-3001	TGS-2201	TGS-2600	AS-MLC	e2v	CCS-801	[24]	[27]	[28]
Sensing film	Tin dioxide (SnO_2)	—	—	Tin dioxide (SnO_2)	—	—	WO_3 μ-hotplate	SnO_2	SnO_2
Targeted gas	Carbon monoxide (CO)	Gasoline (CO, H_2, HC)	Hydrogen (H) and carbon monoxide (CO)	Carbon monoxide (CO)	Carbon monoxide (CO)	Carbon monoxide (CO)	NO_2	CO and NH_3	NO_2
Concentration (ppm)	5–1000	10–1000	1–30	0.5–500	1–1000	10–1000	0–40 ppm 10–100 ppb (detection limit)	0–150 for NH_3	10–30
Sensitivity factor	9 at 100 ppm CO R_{air}/Rs_{gas}	0.35 at 10 ppm H_2 Rs_{gas}/R_{air}	0.3–0.6 at 10 ppm H_2 Rs_{gas}/R_{air}	100 at 30 ppm CO R_{air}/Rs_{gas}	4 at 400 ppm CO R_{air}/Rs_{gas}	4 at 400 ppm CO R_{air}/Rs_{gas}	60 at 40 ppm NO_2 Rs_{gas}/R_{air}	—	—
Operating temperature	—	—	—	270 °C	340 °C	160 °C	200°	270 °C	400 °C
Sensor resistance in air (baseline resistance) Min.	4 MΩ	10 kΩ	10 kΩ	1 kΩ	100 kΩ	0.1 MΩ	200 MΩ	1 kΩ	1 kΩ
Max.	36 MΩ	80 kΩ	90 kΩ	200 kΩ	1.5 MΩ	2 MΩ	20 GΩ	0.1 GΩ	1 GΩ
R_{max}/R_{min}	9×10^0	8×10^0	9×10^0	2×10^2	1.5×10^1	2×10^1	1×10^2	1×10^5	1×10^6
Power consumption	65 mW	505 mW	210 mW	35 mW	76 mW	16 mW	13 mW	450 mW (for heater) Up to 600 mW for interface	30 mW for heater 15 mW for interface

Applied sensor; Cambridge CMOS sensors, CCS801; MiCS, e2v; MSGS, Microsens SA; TGS, Figaro.

prototype sensors such as micromachined sensors, which are more flexible to detect many gases with different concentration levels as requested by versatile gas sensory systems. The sensor electrical response (resistance range variation) corresponds to the variation of a target concentration of common pollutant gases such as Nitrogen dioxide (NO_2) and Carbon monoxide (CO). Table 7.2 reports commercial sensors based on tin Oxide (SnO_2) sensing films for the detection of CO gas. In case of noncommercial (prototype) sensors, the sensors were micromachined sensors, where an embedded heater and thermometer are integrated with the sensor improving the sensing sensitivity. Besides the sensors based on $SnO2$, the performance of a tungsten trioxide (WO_3) sensor fabricated on silicon microhotplate is reported [24]. Table 7.2 also highlights the electrical behavior of MOX-based gas sensor as a sensing resistor of wide range. Depending on the gas concentration range and type, the ratio R_{max}/R_{min} for commercial sensors shows a variation up to two-decades, whereas for micromachined sensors this range might extend up to six decades.

7.3 System Requirements and Literature Review

7.3.1 System Requirements

Table 7.3 reports the wide range resistance variation as an essential metric for the design of the interface circuit. The wide resistance ranges required by indoor and outdoor air monitoring purposes put severe constraints on the electronic interface circuit in terms of linearity and dynamic range (DR). The electronic readout circuit requires a linearity between 0.1 and 1% for air quality monitoring, whereas, a linearity of 10% is adequate for rough detection, such as fire alarms. Moreover, the readout circuit should sustain a DR greater than 140 dB for indoor–outdoor gas monitoring [28, 35].

To summarize, three main specifications define the requirements of the electronic interface circuit, i.e.:

1) Wide range (more than four decades)
2) High linearity (between 0.1 and 1%)
3) Large DR (>140 dB)

The role of the analog electronic readout circuit is to tackle the wide variation of the sensor resistive signal and convert it into a waveform suitable to be easily converted into the digital domain (interfaced through a microcontroller, counter, ADC, etc.) for further signal processing and learning. Integrating all these parts together on one chip leads to a smart gas sensing system.

Table 7.3 Literature review of chemoresistive MOX gas sensors with interface electronics.

		[29]	[30]	[31]	[18]	[32]	[33]	[34]	[35]
	Sensor type	Integrated 1×1 SnO$_2$ sensor	Integrated 4×4 SnO$_2$ sensor array	Integrated 2×2 SnO$_2$ sensors array	Integrated 2×2 SnO$_2$ sensors array	Commercial sensor Figaro TGS 2600	Commercial MG811 sensor	MEMS	Integrated heterogeneous sensor [24]
	Targeted gases	CO, CH$_4$	CH$_4$, H$_2$, ethanol, CO	CO$_2$, CH$_4$, H$_2$O$_3$, ethanol	CO$_2$, CH$_4$, H$_2$O$_3$, ethanol	CO	CO$_2$	—	—
	Range (resistance-frequency)	1 kΩ–10 MΩ	0–20 MΩ	100Ω–20 MΩ	1 kΩ–1 GΩ	100 KΩ–100 GΩ	Frequency range 285–242 kHz	100 Ω–1 MΩ	10 kΩ–10 GΩ
	Application	Environmental monitoring	Environmental monitoring	Environmental monitoring	Environmental monitoring	Environmental monitoring	—	Environmental monitoring	Environmental monitoring
Electronic interface circuit	Architecture	Logarithmic compression	Differential readout circuit	Programmable transresistance amplifier (PTA)	Resistance to frequency conversion	Resistance to period conversion	Resistance controlled oscillator	Resistance to frequency	Resistance to time
	Additional circuits	—	Decoder, voltage shifter, subtractor	Bias circuit, ADC, DAC	Counter	Passive components (resistors, capacitors)	Decoder, 5-bit DAC	Counter	—
	Calibration	No	Yes	Yes	No	No	Yes	No	No
	Technology	0.8 μm–5V CMOS process	5 μm–10V CMOS process	0.35 μm–3.3V CMOS process	0.35 μm–3.3V CMOS process	0.35 μm–1.8V CMOS process	0.35 μm–5V 2P4M CMOS process	0.13 μm standard CMOS technology	0.35 μm–3.3V AMS CMOS technology
	Power consumption	—	15.5 mW	6 mW × 4 channels	15 mW	600 μW	20 mW	450 μA × (1.5V/1.8V)	870 μA from 1.65 V at 10kΩ
	Linearity error	1–2% over 4-decades	—	0.1%	0.8% 1kΩ–100 MΩ	1% over 4-decades	<1.3% after calibration	0.4%	<0.9%
	DR (dB)	115	—	161	141	—	—	128 dB	161
	Silicon area	—	—	3.1 mm^2	0.42 mm^2	0.9 mm^2	1.12 mm^2	125 mm^2	—

7.3.2 Wide Range Resistance Interface Review

Recent efforts in the state-of-the-art aiming to design electronic readout circuitry for interfacing MOX-resistive gas sensors that target the above listed metrics, can be summarized in four viable circuit architectures:

1) Direct measurement
2) Multiscale, autoranging approach
3) Logarithmic compression and
4) Frequency-Time or Duty cycle electronic readout circuit interfaces

Architectures based on direct measurement such as current–voltage conversions, voltage divider [36], or Wheatstone bridge [37, 38] performs direct resistance-to-voltage conversion as a first conversion step followed by an instrumentation or precision differential amplifiers (to reduce offset voltages) in additional to an analog-to-digital convertor (ADC) topology. The main drawbacks of such circuits are: satisfactory accuracy is only attained for small resistance variations, integrating high value resistors in CMOS consumes large chip areas, thus high cost increase.

Multiscale approach [31] and baseline cancelation approach with calibration and conversion [39] are high performance architectures. High linearity and large dynamic range can be achieved over wide resistance range using this architecture. However, the main limitations of this approach are the high hardware cost, the long measurement time, and the increased complexity, due to calibrating the discontinuities among the different scales, which is needed in the digital domain.

Uncalibrated circuits that satisfy the targeted requirements are the logarithmic or quadratic compression approach and the resistance to Time [40, 41]/Frequency [18, 34]/duty cycle conversion [23, 42]. Logarithmic compression-based architectures [29] consist of a voltage to current converter followed by bipolar transistors pair or a pair of diodes. This leads to a logarithmic scale behavior according to the diode equation. This approach suffers from the mismatch between the diodes and the current mirror resulting in limited linearity and DR. Electronic readout circuits based on the resistance to time/frequency/duty cycle conversion architecture, also named quasi-digital [40, 43], or semi digital [44], are considered efficient signal conditioning solutions to readout resistive sensors signals varying over wide range without any need of calibration [18, 23, 40, 41]. Such conversion techniques are advantageous since it is possible to translate the sensed signal into a time/frequency value, which is subsequently digitized and acquired via a simple digital counter. In this case, the digitized output can be easily transmitted to a package of switch networks and radio, infrared, optical, or ultrasound devices [33]. Other advantages are the easiness of multiplexing, the robustness of the sensor output signal with respect to noise, disturbances, and supply voltage or current variations [40, 43]. The hardware cost and complexity are less than that in conventional ADC approaches. Table 7.3 reports efforts done toward the development of a gas sensing system (ASIC/analog readout). The table also reports the performance merits for solutions adapted from literature aiming to tackle wide resistance range with high linearity and large DR.

7.4 Resistance to Time/Frequency Conversion Architecture

7.4.1 Electronic Circuit Description

Resistance to Time (R-to-T) conversion architecture employs a Resistance to current converter (R-to-I) and Current Controlled Oscillator (CCO). The input is an analog signal represented by a resistance value, i.e. the sensor's resistance (R_{sens}) and its output has a period (T) proportion to the input resistance. The voltage across R_{sens} is held fixed to a constant voltage reference (V_{ref}) producing a current (I_{sens}) inversely proportional to R_{sens}, where $I_{sens} = (V_{ref} - V_{ss})/R_{sens}$, V_{ss} is the minimum supply voltage. Then, the current is scaled and mirrored requiring a high precision push/pull current mirror to produce an alternating current variation (I^+, I^-). The alternating current then charges and discharges a capacitor (C), thus converting the current into a voltage signal through a current integrator. Next, the integrator's output is converted to a periodic waveform through the control circuit. The control circuit includes a window comparator that limits the oscillator output swing through comparing the integrator output voltage (Vc(t)) with two reference voltages V_H and V_L (H stands for High and L stands for Low). A square output wave is then generated by an RS-NAND. The RS-Latch forces the comparators to change state alternately, thus, controlling the switches and creating a periodic signal able to provide a digital reading when connected to a counter. The output signal period is proportional to the sensor resistance variation R_{sens}, i.e. $T = k \cdot R_{sens}$, where k depends on several parameters i.e. the comparator's reference voltages V_H and V_L, the integrating capacitor, the voltage across the sensor, as well as β the scaling factor of the current mirror [18, 34, 40].

$$T = \frac{2C\left(V_H - V_L\right)}{\beta \, V_{sens}} R_{sens}$$

Figure 7.2 shows the main blocks of the resistance to time approach, whereas Figure 7.3 represents a circuit scheme of the R-to-T conversion.

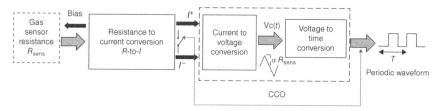

Figure 7.2 Block diagram of the resistance to time conversion.

7.4.2 Specifications for Each Building Block to Preserve High Linearity

To obtain a linearity better than 1% requires to achieve high accuracy in each block. The linearity error (ϵ) is an essential metric since once the R-to-T conversion circuit attains a low linearity error, large DR is resulted, where the DR is defined as [28]:

$$DR = 20\log\frac{1}{\epsilon} \times \frac{\max R_{sens}}{\min R_{sens}}.$$

7.4.2.1 Resistance to Current Conversion (*R*-to-*I*)

Resistance to current conversion requires amplifying the sensor resistive signal at a first step and mirroring the signal through current mirroring topologies.

- Fixing the voltage across the sensor's resistance has the advantage of neglecting the effects of the parasitic capacitance of the sensor, this might be performed using either single ended [18, 34, 45] or differential topologies [22, 42, 43, 46] as reported in Figure 7.4. For the single-ended topology, the voltage is fixed across R_{sens} through a feedback network and an Operational Transconductance Amplifier (OTA) connected to M_{n1} transistor in common source configuration. The OTA requires high DC gain, and low offset [18, 34, 40]. The large open loop gain of the amplifier (>90 dB) forces V_{ref} to appear across R_{sens}, thus $V_{sens} = V_{ref} - V_{ss}$. Thereafter, the input voltage V_{ref} is converted into current by the M_{n1} NMOS transistor (*V*-to-*I*). Considering the differential topology, the voltage is fixed across the sensor using two OTAs such that $V_{sens} = V_{ref1} - V_{ref2}$ in a mode similar to the one of single ended. The voltage across the sensor resistance is bounded between the two reference voltages. The offset of the amplifier highly affects the accuracy of the conversion of the voltage signal into a current one especially when the voltage across the senor is fixed at low values (i.e. mV). An assessment was performed in Ref. [34] to study the effect of the offset voltage on the accuracy when the voltage across the sensor resistance is converted into current. For $V_{sens} = 50$ mV, the study reports that an offset of

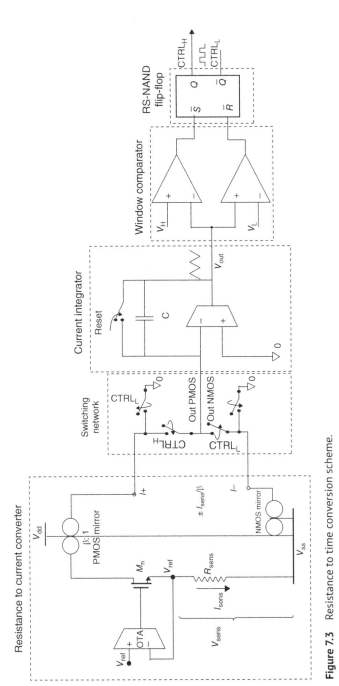

Figure 7.3 Resistance to time conversion scheme.

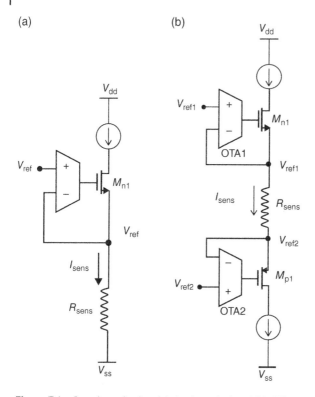

Figure 7.4 *R*-to-*I* topologies: (a) single ended and (b) differential.

200 μV leads to a relative error between −0.5 and 0.5%, whereas, high offset level of 1000 μV results in a relative error between −2.2 and 2.2%. The single ended topology is affected by the supply voltage variation in contrast to differential topology. Differential topology also provides better accuracy, i.e. 0.044% absolute relative error [22] compared with single ended, i.e. 0.26% [45]; however, the voltage headroom in the differential topology is limited by the follower transistor and the number of transistors employed in current mirror.

• The current is mirrored through a high linearity (always relative error below 1%) wide range push/pull architecture. Cascode architectures are used to achieve high accuracy as the current is being delivered to the other subsequent blocks. A possible way to improve the accuracy for low resistance values is to use large transistor dimensions (to keep low overdrive voltage) in parallel configuration as reported in Ref. [35, 45]. Gain boosted current mirrors where a high gain OTA boosts the current mirror output resistance, can also improve the accuracy [40, 45]. Other mirroring topologies are reported in Refs. [18, 22, 34, 42, 43, 46].

7.4.2.2 Switches

The realization of switches as a parallel combination of NMOS and PMOS transistors with dummies rather than the of a single NMOS transistor results in improved precision by minimizing the effect of charge injection. Reference [40] shows the switch network architecture adapted from Ref. [47].

7.4.2.3 Current to Voltage Conversion (*I*-to-*V*)

The wide range specifications put constraints on the integrator as well. The latter should attain a high output voltage swing, and achieve a wide gain bandwidth product. Reference [34] reports for a sensor resistance range of 100–10 MΩ; folded cascade architecture is used achieving a DC gain of 55 dB and a gain bandwidth of 200 MHz when biased with 250 µA. Two stage compensated Miller OTA is adapted in Ref. [35] with a 50 dB DC gain and 200 MHz gain bandwidth product.

7.4.2.4 Voltage to Time/Period (*V*-to-*T*) Conversion

The integrator output is limited between two reference voltages V_H and V_L provided by the window comparator. The window comparator is required to be continuous time, high speed architecture (response time <2 ns) [18, 34, 40, 48]. Each comparator also employs small hysteresis using a positive feedback loop. An RS-NAND flip-flop is connected to the comparator's output resulting in a periodic waveform directly proportional to R_{sens}. The hysteresis and RS-NAND flip-flop forces the switches at the integrator's input to switch alternately guaranteeing proper synchronization of the controlled switches.

Low linearity error over a wide resistance range results in large DR, i.e. 141 dB measured [18], 161 dB simulated [35], and 128 dB simulated [34].

7.5 Power Consumption

7.5.1 Power Consumption of MOX Gas Sensor

MOX gas sensors are considered as power hungry sensors. MOX gas sensors operate at high temperature, typically between 150 and 500 °C. To insure fast and reversible operation, heating elements are integrated within the sensor resulting in high power consumption in hundreds of mW for the sensor only. MOX gas sensors integrated on microhotplates (µ-hotplate) instead of alumina substrate, decrease the power consumption down to tens or even several mW. Table 7.2 highlighted the power consumption levels for MOX sensors integrated on µ-hotplate [24] and others integrated on alumina or silicon substrate [27]. More interestingly, to reduce the operating temperature and to enhance surface chemistry, new techniques were tested causing additional reduction in power consumption. Such techniques include modifying the surface of the sensing layer with noble metal nanoparticles, the use of electrostatic field, and of UV activation of reactants [49]. Self-heated MOX nanostructures [50] have been used for ultra-low power gas sensors. Reference [49] also addresses, in detail, the techniques and fabrication process used toward reducing the power consumption aiming to develop an ultra-low power consumption MOX gas sensor for mobile applications.

7.5.2 Low Power Operating Mode

Another possible way to overcome the power consumption limitation of MOX gas sensors is to control its heater, which is the main cause of power consumption, at circuit level. This means to reduce the sensor's heating time through determining air quality roughly from the very beginning of the sensor's transient response [51], thus avoiding the continuous heating of the MOX gas sensor and saving power. Strategies based on duty cycle have also been addressed in literature for commercial MOX gas sensors allowing a great increase in the battery lifetime for battery powered embedded electronic devices [52, 53].

7.5.3 Power Consumption at Circuit Level

Once the power consumption of MOX gas sensor is reduced and optimized for battery operated portable devices, reducing the power consumption at circuit level should be noticed as well. Considering R-to-T circuit architecture, some proposed and adapted techniques from literature, to design low power consumption electronic readout circuit, are discussed below.

The main cause of high power consumption at circuit level for R-to-T circuits, is when interfacing low resistance values, i.e. hundreds of Ohms to hundreds of kilo-Ohms. Starting with the R-to-I conversion, a possible solution is to design a low reference voltage when fixing the voltage across the sensor. This results in a lower current consumption since $I_{sens} = (V_{ref} - V_{ss})/R_{sens}$. However, designing a low reference voltage can affect the linearity of the circuit especially when high resistances are to be considered. Different input reference voltage levels can be a solution. Different voltage ranges can be assigned to each resistance variation range through a switching network [31]. However, a sort of calibration to control the switches of each voltage level should be noticed in the digital domain.

Flipped voltage follower (fvf) is a basic cell for low power, low voltage applications [54]. A proposed idea is to adapt fvf in the R-to-I instead of a source follower transistor. The low output resistance of the fvf improves the driving capability especially when low sensor resistances are to be detected at which the current consumption is high. Whereas, for MOX sensors with R_{sens} range $>100\,k\Omega$, the current biasing in the integrator, as well as the comparators can be reduced more than 10 times compared to including lower resistance ranges, highly reducing the current and thus the power consumption of the circuit.

7.6 Conclusion

In this chapter, we introduced MOX based semiconductor gas sensors as good candidates for realizing a gas sensing system due to their variety of advantages such as low cost, long life time, high sensitivity, small size, simplicity of operation, flexibility in production, and compatibility with the standard CMOS process. A review of MOX gas sensors (commercial and noncommercial) was presented to draw the main requirements for the electronic interface readout circuit design. The chapter presented a review of the state-of-the-art for possible circuit interfaces provided in literature. Electronic interface circuits based on time/frequency are preferable when converting wide resistance range with high linearity (low linearity error) without any need of calibration since the output periodic waveform can be easily converted to digital domain. A detailed explanation for each building block constituting R-to-T architecture has been provided highlighting the main requirements to achieve high linearity and wide DR.

One of the top design requirements for semiconductor chemical sensors is to develop low cost, simple, reliable, and sensitive sensors to be used in handheld portable devices, e.g. smart phones, computers, tablets, and wearable devices [55]. This chapter discussed the efforts done toward reducing the power consumption which is the main barrier preventing the use of MOX gas sensor for mobile applications. The trend in modern electronics is to integrate several sensors with different types (gas, pressure, temperature, humidity, touch, etc.) in a single device. Time/frequency-based electronic circuit was recently introduced in literature as an efficient circuit architecture interfacing resistive and capacitive sensors types [56]. Digital signal processing and integrating intelligence represent another research domain, toward developing a compact optimized gas sensing system, to be addressed for future work.

References

1 David A. Bengston, Diane S. Henshel, "Environmental Toxicology and Risk Assessment: Biomarkers and Risk Assessment", ASTM International, 1996, ISBN 0803120311, p 220.
2 Harwood, D. (2001). Something in the air [electronic nose]. *IEEE Review* 47: 10–14.
3 Zhang, L., Tian, F., Nie, H. et al. (2012). Classification of multiple indoor air contaminants by an electronic nose and a hybrid support vector machine. *Sensors and Actuators B: Chemical* 174: 114–125.
4 Dragonieri, S., van der Schee, M.P., Massaro, T. et al. (2012). An electronic nose distinguishes exhaled breath of patients with malignant pleural mesothelioma from controls. *Lung Cancer* 75: 326–331.

5 Guo, D., Zhang, D., Li, N. et al. (2010). A novel breath analysis system based on electronic olfaction. *IEEE Transactions on Biomedical Engineering* 57: 2753–2763.

6 Wilson, A.D. and Baietto, M. (2011). Advances in electronic-nose technologies developed for biomedical applications. *Sensors* 11: 1105–1176.

7 Haddi, Z., Amari, A., Alami, H. et al. (2011). A portable electronic nose system for the identification of cannabis-based drugs. *Sensors and Actuators B: Chemical* 155: 456–463.

8 Di Natale, C., Macagnano, A., Martinelli, E. et al. (2001). The evaluation of quality of post-harvest oranges and apples by means of an electronic nose. *Sensors and Actuators B: Chemical* 78: 26–31.

9 Concina, I., Falasconi, M., and Sberveglieri, V. (2012). Electronic noses as flexible tools to assess food quality and safety: should we trust them? *IEEE Sensors Journal* 12: 3232–3237.

10 Macías, M., Manso, A., Orellana, C. et al. (2013). Acetic acid detection threshold in synthetic wine samples of a portable electronic nose. *Sensors* 13: 208–220.

11 Ampuero, S. and Bosset, J.O. (2003). The electronic nose applied to dairy products: a review. *Sensors and Actuators B: Chemical* 94: 1–12.

12 Berna, A. (2010). Metal oxide sensors for electronic noses and their application to food analysis. *Sensors* 10: 3882–3910.

13 Baldwin, E.A., Bai, J., Plotto, A., and Dea, S. (2011). Electronic noses and tongues: applications for the food and pharmaceutical industries. *Sensors* 11: 4744–4766.

14 Hasan, N., Ejaz, N., Ejaz, W., and Kim, H. (2012). Meat and fish freshness inspection system based on odor sensing. *Sensors* 12: 15542–15557.

15 Gardner, J.W., Shin, H.W., Hines, E.L., and Dow, C.S. (2000). An electronic nose system for monitoring the quality of potable water. *Sensors and Actuators B: Chemical* 69: 336–341.

16 Baby, R.E., Cabezas, M., and Walsöe de Reca, E.N. (2000). Electronic nose: a useful tool for monitoring environmental contamination. *Sensors and Actuators B: Chemical* 69: 214–218.

17 Ho, C. and Hughes, R. (2002). In-situ chemiresistor sensor package for real-time detection of volatile organic compounds in soil and groundwater. *Sensors* 2: 23–34.

18 Grassi, M., Malcovati, P., and Baschirotto, A. (2007). A 141 dB dynamic range CMOS gas-sensor interface circuit without calibration with 16-bit digital output word. *Journal of Solid-State Circuits* 42: 1543–1554.

19 Goschnick, J. and Harms, M. (2002). Landmine detection with an electronic nose mounted on an airship. *NATO Science Series* 66: 83–91.

20 Wilson, A.D. (2012). Review of electronic-nose technologies and algorithms to detect hazardous chemicals in the environment. *Procedia Technology* 1: 453–463.

21 Tsow, F., Forzani, E., Rai, A. et al. (2009). A wearable and wireless sensor system for real-time monitoring of toxic environmental volatile organic compounds. *IEEE Sensors Journal* 9: 1734–1740.

22 Hijazi, Z., Caviglia, D., Chible, H., and Valle, M. (2016). Differential *R*-to-*I* conversion circuit for gas sensing in biomedical applications. In: *2016 3rd Middle East Conference on Biomedical Engineering (MECBME), Beirut*, 76–79. Piscataway, NJ: IEEE.

23 Tang, K.T., Chiu, S.W., Chang, M.F. et al. (2011). A low-power electronic nose signal-processing chip for a portable artificial olfaction system. *IEEE Transactions on Biomedical Circuits and Systems* 5 (4): 380–390.

24 Barborini, E., Leccardi, M., Bertolini, G. et al. (2007). Batch fabrication of nanostructured heterogeneous microarrays for chemical sensing. In: *NSTI Nanotech: Nanotechnology Conference and Trade Show, Santa Clara, CA (20–24 May 2007): Techical Proceedings*, vol. 1, 336–338. CRC Press.

25 Gardner, J.W., Guha, P.K., Udrea, F., and Covington, J.A. (2010). CMOS interfacing for integrated gas sensors: a review. *IEEE Sensors Journal* 10 (12): 1833–1848.

26 Liu, H., Zhang, L., Li, K.H.H. et al. (2018). Microhotplates for metal oxide semiconductor gas sensor applications – towards the CMOS-MEMS monolithic approach. *Micromachines* 9 (11): 557. https://doi.org/10.3390/mi9110557.

27 Sharma, D.K., Dwara, R.S.V., Botre, B.A. et al. (2017). Temperature control and readout circuit interface for MOX based NH_3 gas sensor. *Microsystem Technologies* 23: 1575–1583. https://doi.org/10.1007/s00542-016-3126-6.

28 Malcovati, P., Grassi, M., and Baschirotto, A. (2013). Towards high-dynamic range CMOS integrated interface circuits for gas sensors. *Sensors and Actuators B: Chemical* 179: 301–312.

29 Barrettino, D., Graf, M., Hafizovic, S. et al. (2004). A single-chip CMOS micro-hotplate array for hazardous-gas detection and material characterization. In: 2004 IEEE International Solid-State Circuits Conference (IEEE Cat. No.04CH37519), 312–313. IEEE.

30 Guo, B. and Bermak, A. (2005). A differential readout circuit for tin oxide gas sensor array. In: *2005 IEEE Conference on Electron Devices and Solid-State Circuits, Howloon, Hong Kong (19–21 December 2005)*, 635–638. IEEE.

31 Grassi, M., Malcovati, P., and Baschirotto, A. (2007). A 160 dB equivalent dynamic range autoscaling interface for resistive gas sensors arrays. *Journal of Solid-State Circuits* 42: 518–528.

32 DeMarcellis, A., Depari, A., Ferri, G. et al. (2013). A CMOS integrated low-voltage low-power time-controlled interface for chemical resistive sensors. *Sensors and Actuators B: Chemical* 179: 313–318.

33 Chiang, C.T., Chung, M., and Huang, M.Y. (2016). Design of a gas sensor transducer circuitry with calibration ability for CO_2 concentration detection. *IEEE Sensors Journal* 16 (16): 6367–6373.

34 Ciciotti, F., Baschirotto, A., Buffa, C., and Gaggl, R. (2017). A MOX gas sensors resistance-to-digital CMOS interface with 8-bits resolution and 128 dB dynamic range for low-power consumer applications. In: *2017 13th Conference on Ph.D.*

Research in Microelectronics and Electronics (PRIME), Giardini Naxos (12–15 June 2017), 21–24. IEEE.

35 Hijazi, Z., Grassi, M., Caviglia, D.D., and Valle, M. (2018). Time-based calibration-less read-out circuit for interfacing wide range MOX gas sensors. *Integration* 63: 232–239.

36 Shurmer, H.V. and Gardner, J.W. (1992). Odor discrimination with an electronic nose. *Sensors and Actuators B: Chemical* 8 (1): 1–11.

37 Cole, M., Gardner, J.W., Lim, A.W.Y. et al. (1999). Polymeric resistive bridge gas sensor array driven by a standard cell CMOS current drive chip. *Sensors and Actuators B: Chemical* 58 (1–3): 518–525.

38 Leung, C.K. and Wilson, D.M. (2005). Integrated interface circuits for chemiresistor arrays. In: *2005 IEEE International Symposium on Circuits and Systems, Kobe, Japan (23–26 May 2005)*, 5914–5915. IEEE.

39 Lin, Y., Gouma, P., and Stanacevic, M. (2013). A low-power wide-dynamic-range readout IC for breath analyzer system. In: *2013 IEEE International Symposium on Circuits and Systems (ISCAS 2013), Beijing*, 1821–1824. Piscataway, NJ: IEEE.

40 Hijazi, Z., Grassi, M., Caviglia, D.D., and Valle, M. 153 dB dynamic range calibration – less gas sensor interface circuit with quasi – digital output. In: *2017 New Generation of CAS (NGCAS), Genova (6–9 September 2017)*, vol. 2017, 109–112. IEEE https://doi.org/10.1109/NGCAS.2017.11.

41 DeMarcellis, A., Depari, A., Ferri, G. et al. (2008). Uncalibrated integrable wide-range single-supply portable interface for resistance and parasitic capacitance determination. *Sensors and Actuators B: Chemical* 132 (2): 477–484.

42 Gómez-Ramírez, E., Maeda-Nunez, L.A., Álvarez-Simón, L.C., and Flores-García, F.G. (2019). A highly robust interface circuit for resistive sensors. *Electronics* 8: 263.

43 Azcona, C., Calvo, B., Celma, S. et al. (2013). Ratiometric voltage-to-frequency converter for long-life autonomous portable equipment. *IEEE Sensors Journal* 13 (6): 2382–2390.

44 Lu, J.H.L., Inerowicz, M., Joo, S. et al. (2011). A low-power, wide-dynamic-range semi-digital universal sensor readout circuit using pulsewidth modulation. *IEEE Sensors Journal* 11 (5): 1134–1144.

45 Hijazi, Z., Caviglia, D., Valle, M., and Chible, H. (2016). Wide range resistance to current conversion circuit for resistive gas sensors applications. In: *2016 12th Conference on PhD Research in Microelectronics and Electronics (PRIME), Lisbon*, 1–4. IEEE.

46 Conso, F., Grassi, M., Malcovati, P., and Baschirotto, A. (2012). Reconfigurable integrated wide-dynamic-range read-out circuit for MOX gas-sensor grids providing local temperature regulation. In: *SENSORS, 2012, Taipei (28–31 October 2012)*, 1822–1825. IEEE.

47 Razavi, B. (2000). *Design of Analog Integrated Circuits*. McGraw-Hill.

48 Koay, K.C. and Chan, P.K. (2015). A low-power resistance-to-frequency converter circuit with wide frequency range. *IEEE Transactions on Instrumentation and Measurement* 64 (12): 3173–3182.

49 Jang, H.W., Choi, Y.R., and Kim, Y.H. Novel metal oxide gas sensors for mobile devices. In: *Smart Sensors and Systems*, 131–153. Cham: Springer.

50 Strelcov, E., Dmitriev, S., Button, B. et al. (2008). Evidence of the self-heating effect on surface reactivity and gas sensing of metal oxide nanowire chemiresistors. *Nanotechnology* 19: 355502.

51 Jelicic, V., Oletic, D., Sever, T., and Bilas, V. (2015). Evaluation of MOX gas sensor transient response for low-power operation. In: *2015 IEEE Sensors Applications Symposium (SAS), Zadar (13–15 April 2015)*, 1–5. IEEE.

52 Rossi, M. and Brunelli, D. (2014). Ultra low power MOX sensor reading for natural gas wireless monitoring. *IEEE Sensors Journal* 14 (10): 3433–3441.

53 Burgués, J. and Marco, S. (2018). Low power operation of temperature-modulated metal oxide semiconductor gas sensors. *Sensors* 18: 339.

54 Ramirez-Angulo, J., Ducoudray-Acevedo, G., Carvajal, R.G., and Lopez-Martin, A. (2005). Low-voltage high-performance voltage-mode and current-mode WTA circuits based on flipped voltage followers. *IEEE Transactions on Circuits and Systems II: Express Briefs* 52 (7): 420–423.

55 Chou, T.-I., Chang, K.-H., Jhang, et al. (2018). A 1-V 2.6-mW environmental compensated fully integrated nose-on-a-chip. *IEEE Transactions on Circuits and Systems II: Express Briefs* 65 (10): 1365–1369.

56 Ciciotti, F., Buffa, C., Gaggl, R., and Baschirotto, A. (2018). A programmable dynamic range and digital output rate oscillator-based readout interface for MEMS resistive and capacitive sensors. In: *2018 International Conference on IC Design & Technology (ICICDT), Otranto (4–6 June 2018)*, 41–44. IEEE.

8

Smart and Intelligent E-nose for Sensitive and Selective Chemical Sensing Applications

Saakshi Dhanekar

Centre for Biomedical Engineering (CBME), Indian Institute of Technology (IIT), New Delhi, India
Department of Electrical Engineering, Indian Institute of Technology Jodhpur, Karwar, Rajasthan, India

8.1 Introduction

For several decades hitherto the fields of robotics and artificial human-like machines have been emerging and evolving. Even now this field has a lot to explore both theoretically and experimentally. There is a great zest in making human-like machines/devices as these can be used in harsh conditions where human survival is a challenge. This can be at very high and low temperatures or where there are space constraints or very low gravity. The sensors used in such machines can be based on electrical, optical, piezoelectric, mechanical, electrochemical, gas chromatography (GC), mass spectrometry (MS), or a combination of these. These sensors are trained through data and learning so they acquire and build a basic capability to think and decide. One such a device or a system is e-nose which is known to be an intelligent system which senses the presence of analytes around it [1–3]. It then makes a decision and identifies a specific vapor among other analytes. Such a system comprises sensors array, necessary circuitry, and signal processing. Thus, it requires expertise in materials, sensors, circuits, signal processing, and data analysis. Over the years, e-noses have provided a plethora of benefits in various applications [4–8].

Have you wondered why only smell remains the most crucial parameter? Because the freshness and quality of food can be quickly assessed through smell. All products like meat, fish, honey, fruits, vegetables, and dairy products allow to get characterized through their aroma. This can help in assessing their

Smart Sensors for Environmental and Medical Applications, First Edition. Edited by
Hamida Hallil and Hadi Heidari.
© 2020 The Institute of Electrical and Electronics Engineers, Inc.
Published 2020 by John Wiley & Sons, Inc.

maturity stage, microbial contamination, adulteration, etc. It is known that the smell is subjective and every human has a different degree of reaction to all smells. Thus, smell recognition is entirely based on personal experience and ability to understand smells. Human olfactory system is the most natural and known system by the human community for identification of various odors and flavors. Want to know how it works? It is simple to understand as this can be related to our human body.

8.1.1 The Human Olfactory System

In humans, there are different sensory systems which are part of the sense: (i) olfaction (sense of smell), (ii) gustation (sense of taste), (iii) vision (sense of seeing), (iv) auditory (sense of hearing), etc. [3]. The human olfactory system is supposed to be the most dominant factor to decide on the type of flavor. This system aides in odorants to bind to the specific sites in the olfactory systems on the receptors. There are around 10 million olfactory cells and around 400 different olfactory receptors [9]. The olfactory structure consists of sensory neurons in the olfactory epithelium which captures odorant molecules and translates these into signals which reach the olfactory bulb and then the brain. Thus, the molecules go through the nasal passage and dissolve in the mucus which secrets odorant-binding proteins. The odor gets diffused by binding to this protein and is detected by the olfactory receptors in the sensory neurons. The olfactory bulb is in vertebrate forebrain and helps in sense of smell [10]. A schematic of the same is shown in Figure 8.1.

8.1.2 The Artificial Olfactory System

This kind of system works electronically and mimics the human nose. The concept of e-nose was first introduced by Persaud and Dodd [12] and Ikegami and Kaneyasu [13] in 1982 which consisted of nonspecific sensors arranged in the form of array and a patterns recognition technique. In 1988, the term "e-nose" was defined by Gardner and Bartlett as "an instrument which comprises of an array of electronic chemical sensors with partial specificity and appropriate pattern recognition system, capable of recognizing simple or complex odors" [3, 14]. Main interest in e-nose in the area of biosensors rose around 1991 during a session on artificial olfactory in the workshop of the North Atlantic Treaty Organization (NATO). E-nose is a very useful tool to identify and distinguish between complex and simple smell. There is a sensor array and different pattern recognition techniques used in the e-nose system which are discussed in this section. An introductory figure to explain human and artificial olfactory system is shown in Figure 8.2. This explains a comparison of functioning of human brain with artificial system with every stage of aroma detection.

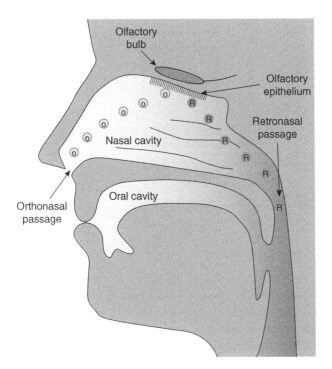

Figure 8.1 Schematic of human olfactory system. The odorants labeled "O" travel to the olfactory epithelium through the nostrils, or orthonasal pathway; the odorants labeled "R" travel through the nasopharynx, or retronasal passage, to the olfactory epithelium. Odorant molecules interact with the olfactory epithelium to generate a signal that is transmitted to the brain via the olfactory bulb. The three horizontal lines represent the turbinates which provide turbulence and mixing to air in the nose Adapted from [11].

8.1.2.1 Sensor Array

The sensors used in the array are nonspecific and give response to vapors differently. The properties of these sensors respond to the change in the ambient. The properties could be any of the physical property of the material. With the advent in the nanotechnology, the sensing response of the materials has touched heights [15–19]. The most common type of sensors used in an array are metal oxide (MOX) based. These consist of a metal core and a boundary of oxygen species around it, between two grains, it is called as grain boundary. Upon exposure to the vapors, adsorption takes place, which, depending on the type of vapor (reducing/oxidizing), increases or decreases the oxygen species [20]. This leads to a change in the intergrain potential barrier height, resulting in change in resistance of the sensitive layer. This resistance is a kind of physical parameter of the MOX-based sensor. A lot of research has happened in the recent past for developing such

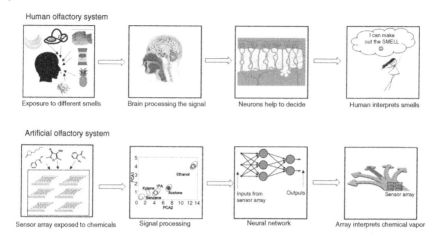

Figure 8.2 Human vs. artificial olfactory system.

sensors. These sensors are very sensitive; however, they operate at high temperatures [21–23], which makes their power consumption very high and increases the fabrication complexity. This problem even piles up when sensors are used in an array. Thus, sensors with high sensitivity and room temperature operation are likely to have best suitability in an array. Such room temperature operation can be achieved by playing with physics of materials and tuning their band gaps. This can be made possible by using heterostructures [24] with near band gaps and literature has evidenced that this dissimilarity has been taken as an advantage in sensing applications. Dwivedi et al. deposited TiO_2 on porous silicon and tested the sensor for the presence of volatile organic compounds (VOCs) [25]. A high response to ethanol was noticed among other analytes at room temperature. Another example of such a sensor is by employing MoO_3 on nano-silicon [26]. This could detect acetone vapors to sub ppm level without heating the sensor. Few other reports have shown integration of micro-electro-mechanical systems (MEMS) to be a solution for reducing the power consumption of the sensor, in cases where heating is required to obtain optimum response. In one of the examples, a MEMS heater was integrated with a heterostructure to obtain the optimum response at around 100 °C at lesser voltage [27].

8.1.2.2 Multivariate Data Analysis

This analysis consists of set of techniques useful in reducing the dimensions in a multivariate problem so that the variables (which are partly correlated) can be displayed in two or three dimensions [4, 28]. One of the most common techniques in multivariate data analysis (MDA) is Principal Component Analysis (PCA) which is a mathematical process that transforms a number of correlated variables into a number of uncorrelated variables called principal components. The others include Canonical Data Analysis, Featured Within and Cluster Analysis.

8.1.2.3 Pattern Recognition Methods

Back-Propagation Neural Networks (BPNN) It is a multilayer feed-forward network trained according to error back propagation algorithm. This has an input, hidden, and output layer, whose mapping relations are described by mathematical equations. A basic structure of BPNN system is shown in Figure 8.3a [30]. First part of the model is forward calculations from input to hidden and to the output layer. In case, a misinformation gets conveyed to the output layer then it is back propagated and the weights of each neuron is recalculated. After the weights are corrected, the information is sent again with less error. The response matches the desired response through repeated iterations.

Self-organization Models It is an artificial neural network (ANN) based on unsupervised learning methodology. There is an input layer and a completion layer. The first layer has the input data and then through order weighing the output

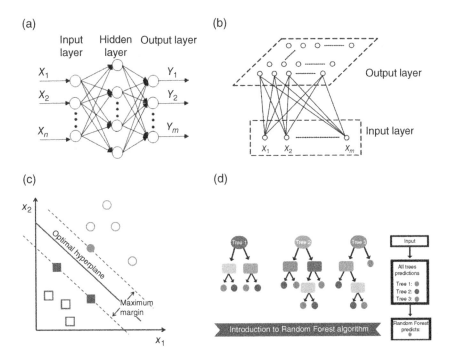

Figure 8.3 Examples of pattern recognition technique: (a) Back-Propagation Neural Networks (BPNN) [Adapted from 30], (b) self-organization models [Adapted from 31], (c) Support Vector Machine (SVM) [Adapted from 29] and (d) Random Forest algorithm [Adapted from 33].

numerical value is given at the output layer. The basic structure of the SOM network is shown in Figure 8.3b [31].

Support Vector Machines It is a supervised learning model with associated learning and regression analysis. It is a discriminative classifier determined by a separating hyperplane. This hyperplane in 2D is a lane dividing a plane in two sections where the classes are separated in either side. In case of nonlinear separable classification, a kernel function is applied to transform original space into a higher-dimensional space. The most common kernels are linear, polynomial, radial basis function, and sigmoid [32]. An example is illustrated in Figure 8.3c.

Random Forest As the name says, it comprises a decision making process by creating different multiple decision trees at different training times. The higher the number of trees, the more the accuracy of the results. The output is the mode of classification and the mean prediction of the individual trees. A model of the same is shown in Figure 8.3d.

Extreme Gradient Boosting (XGBoost) It is a highly effective and widely used machine learning technique. It is known to be a scalable system for tree boosting designed for speed and performance. Additive training strategies are applied for "boosting" to make weak learners as strong learners [32].

8.2 What Is an Electronic Nose?

The most important component of an e-nose is a sensor array comprising of independent sensors giving response to vapors differently. It also consists of an information processing unit, a software for pattern recognition, and reference data [34–36]. A sensor is a device which gives response to stimuli which could be caused by a change in the electric current flow, mechanical stress, bioinduced reaction, or color in the substrate. A sensor alone only shows a change in the properties upon exposure to analytes and converts any type of energy into electrical. A transducer or an actuator converts electrical signal to nonelectrical energy. It is a part of a more complex sensors for example, a loudspeaker which converts electrical energy into magnetic field and then into acoustic waves. The sensor array in an e-nose generates data from all the sensors. This data is processed and a pattern is generated after seeing the trend of the response of each sensor. Thus, the data itself becomes a learning platform for the array. The identity of a simple or complex mixture is identified by a unique pattern generated as a response to these vapors. There is no requirement to physically separate the vapors. The learning is done through a neural network technique which looks for difference between the patterns of the response to different vapors including the reference library. The process is carried out till a selected discrimination is established. This is also recorded in the reference data (already fed) and the unknown data is compared to it [4]. Thus, this system becomes an artificial intelligence trained e-nose system. Several prototypes of e-nose have been developed to solve complex mixtures containing different VOCs [37, 38]. Many sensor types depending on either the technology or material are the building blocks of these e-noses. These include materials like porous [39], MOX [40], conductive electroactive polymers [41, 42], polymers [43], etc.

8.3 Applications of E-nose

8.3.1 Key Applications of E-nose

There are numerous fields where specific detection is the prime concern. To name a few application areas of e-nose are food quality assessment, agriculture, environmental monitoring, packaging, drug, automobile, disease diagnosis, quality control of products, scents, and perfumes. Table 8.1 shows the applications and how e-nose can be helpful in sensing.

8.3.2 E-nose for Chemical Sensing

Conventionally, e-nose was found to be applicable in chemical sensing and over the years it started creeping in biohealthcare and other areas. Aroma detection can be very significant in identifying the presence of gas/vapors. This can be extended to alcohol detection, diseases diagnosis, food quality, etc. In human beings, the olfactory system is the primary mode for identification of aroma. Aroma comes from the aromatic compounds mainly known because of their structural make, functional groups, and chemical bonds. The volatility of these compounds is determined by their bond strengths and polarity. The main cause of aroma is the functional groups which adds a particular smell to the food. The American society for Testing and Materials (ASTM) has defined 836 aroma descriptors [56]. As per the literature, human olfactory can identify less than 100 aromas after good training [57]. The presence of aroma is based on four parameters: threshold, intensity, quality, and hedonic assessment [4]. Threshold means the minimum concentration of aroma that can be sensed Intensity is the strength of aroma detection. Quality is linked to the description of aroma (type of fragrance). The hedonic assessment pertains to the pleasantness or unpleasantness of

Table 8.1 Key applications of e-nose.

Application areas	Details
Food quality assessment [44, 45]	Classification of samples of poultry meat on the basis of shelf-life, and also samples of rapeseed oil based on the extent of thermal degradation, also detection of adulterations of extra virgin olive oil with rapeseed oil, consumer fraud prevention, ripeness, food quality, using aroma sensor
Medical and clinical diagnosis [46]	Rapid detection of tuberculosis (TB), *Helicobacter pylori* (HP) and urinary tract infections (UTI), pathogen detection
Hazardous chemicals and explosives sensor [47], military	Detection of chemicals like benzene, toluene, ethyl benzene, xylene, and gasoline
Chemical sensor [48]	Detection of hydrogen, ethanol, and nitrogen dioxide
Agriculture and forestry [49]	Environmental monitoring, crop protection, plant production, wood and paper processing, forest health protection, and waste management
Cosmetics [50]	Fragrance additives, perfumes quality improvement
Environment [51]	For environmental pollution, detection of microbial toxins, residential organic compost piles
Biomedicine [52]	Detection of human diseases through microbial pathogens, pharmacology, physiology, remote healthcare, wound healing
Risk assessment [53]	Volcanic gases like fumarolic gases detection
Air quality assessment [54]	Temperature, humidity, air quality analysis
Alcohol breath analyzer [55]	Use of sensor array to discriminate alcohol among other VOCs
Scientific research [4]	Botany, pathology, material properties, ANN technique

the aroma. This aroma is a chemical vapor which could be an industrial pollutant, environmental gas/vapor, or an element in a breath sample. It is well known that human breath has many chemical vapors, which are indicators of human health. Few advantages of detection of vapors in breath using e-nose would be: (i) Noninvasive sensing and thus requires no harm to the human body, (ii) detection of a wide variety of vapors or diseases (by generating a pattern), (iii) choice of sensors is based on the type of detection, and (iv) levels of detection can be tuned by small changes in training of the system.

One of the studies shows that the information-theoretic feature selection method (specific to the problem) with machine learning techniques and a subset of E-nose measurement features can be used in quantitative assessment (cross-validated) of the vapors in breath [58].

8.4 Types of E-nose

There are several companies which are into manufacturing of e-nose systems. Since, their applications are so vast, their manufacturing is also rapidly growing. Few of the products with names of their manufacturers and applications are given in Table 8.2.

Table 8.2 Types of e-nose [59].

Manufacturer	Product	Application	Consists of/Uses
The eNose Company, Rotterdam, The Netherlands	AEONOSE	Disease detection using exhaled human breath [60]	Three microhotplate metal oxide sensors; these are heated between 260 and 340 °C in 32 steps during exposure to exhaled breath [61]
Airsense Analytics GmnH, Schwerin, Germany	PEN3	Food, wine quality check [62]	10 various metal oxides consisting of single thick film sensors working at temperatures ranging from 350 to 500 °C [63]
Airsense Analytics GmnH, Schwerin, Germany	Olfosense	Air quality monitoring [64]	A combination of three types of sensors: photoionization detector (PID), electrochemical cell (EC), four metal oxide semiconductors (MOS)
Alpha-Mos, Toulouse, France	HERACLES	Quality control [65]	Flash gas chromatography technology
Sensigent, Baldwin Park, CA, USA	CYRANOSE 320	Medical, food quality assessment [66]	An array of 32 polymer carbon black composite sensors [67]
Electronic Sensor Technology Inc., Newbury Park, CA, USA	Z-Nose	Investigation, food, environment [68]	Surface acoustic wave sensor technology [68]
Owlstone Inc., Cambridge, UK	LONESTAR	Food industry, materials, industry in general [70]	Field asymmetric ion mobility spectrometry (FAIMS) – it is a way of characterizing ionized VOCs according to the difference in drift through a buffer gas under the influence of an oscillating electric field [71]
U-BIOPRED	Comon-Invent eNose	VOCs [72]	Eight MOX sensors, resistance change [71]
The Sensors Group at the Tor Vergata University in Rome	TEN 2010	Biomedical, breath analysis [73]	Quartz crystal microbalance (QCM) covered with metalloporphyrins [71]

8.5 Examples of E-nose

There have been several attempts to classify odors and smells; however, it is very challenging to make an e-nose which is simple, accurate, reliable, and sensitive. One of the early examples of an e-nose was MOSES-II by GSG [74]. It consisted of different MOX- and Quartz crystal microbalance (QCM)-based sensors and was a precise system; however, it faced issues with portability. E-nose also appeared as X-am 7000 by Dräger [75] or the MultiRAE by RAE Systems [76] which were designed for very specific applications. Another interesting electronic nose setup with versatility was introduced as a prototype with ten sensors. This e-nose was used for detection of volcanic gases like fumarolic gases within Solfatara crater [53]. Sanchez-Garrido and team published some work explaining the architecture of an e-nose system and it additionally explains the spatiotemporal analysis of gases [77]. A prototype was built for food quality assessment which used an array of electrochemical sensors and SVM method has been described by Wojnowski et al. [78]. A pneumatic assembly consisting of four electrochemical sensor ch5mbers was used with microprocessors and other necessary circuitry. The picture of the prototype is shown in Figure 8.4. The sensors used were commercially available and gave higher response to specific gases. These also experienced cross-sensitivity like many other gas sensors and so their data was combined and the system was trained using SVM algorithm. This helped in classifying samples of

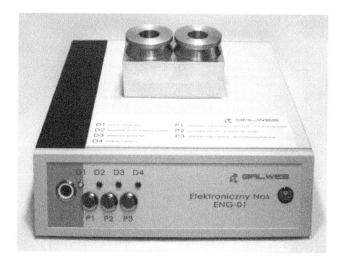

Figure 8.4 The view of an electronic nose prototype based on the electrochemical sensor array Adapted from [45].

poultry meat and of rapeseed oil based on the extent of thermal degradation based on shelf-life with 100% accuracy. It also detected the debasements of extra virgin olive oil with rapeseed oil with 82% overall accuracy.

E-nose has been used to check the mandarin monitoring for crop monitoring. The VOCs from mandarin were collected and measured through a device called as PEN2. The data analysis for e-nose is implemented through PCA and linear discriminant analysis (LDA) method (Figure 8.5) to investigate the capability of the e-nose to identify among different picking-date. This work successfully showed the higher concentration of VOCs from plants attacked by insects or with disease [79]. It also displayed that the efficiency of LDA was much higher than PCA.

Semiconductor MOXs have a wide range of sensing response to different VOCs. E-nose is a device which helps in countering this issue by training the system with data collected in the past. A template with four sensors based on MOX was built; these were mainly In_2O_3 nanowires, SnO_2 nanowires, ZnO nanowires, and single-walled carbon nanotubes (SWNT). One discrimination factor was use of p- and n-type SWNT. The other was micromachine-enabled heating platforms for each sensor. The morphological analysis of four different sensors is given in Figure 8.6 [48, 80]. These sensors were exposed to different gases at various concentrations and at different temperatures to get a lot of data for training the system. PCA was done for discrimination of chemicals (gases) and this was achievable with great accuracy.

Another work [55] illustrates the development of sensor chips from scratch, their material and electrical characterization, response by these chips to VOCs, and how these can be arranged in an array and ethanol can be discriminated from the mixture of vapors. The process to fabricate the sensors is repeatable and scalable making the device prepared for batch fabrication ready for industry production (Figure 8.7a–e). This discusses three sensors: only porous silicon, only TiO_2, and TiO_2 on porous silicon. PS was formed using electrochemical anodization technique. A thin layer of TiO_2 was sputtered on PS and crystalline silicon using RF reactive sputtering. These were later annealed for attaining anatase phase of TiO_2. Figure 8.7f shows the resistive model of TiO_2/PS heterostructure-based sensor. It can be seen that there are mainly three material resistances (R_{TiO_2}, R_{PS}, R_{c-Si}) and two interface resistances, labeled as $R_{TiO_2/PS}$ and $R_{PS/c-Si}$. The interface resistance $R_{TiO_2/PS}$ was found to have played a significant role in sensing mechanism in comparison to single material resistances. Figure 8.7g depicts a 3D schematic of the fabricated sensor. A detailed characterization was done and SEM study reveals a porous morphology of TiO_2/PS with well joined pores formed into tracks of around 50 nm in width and the underlying PS layer bearing pores of 4–6 nm in diameter. Raman and XRD confirmed the structural make of the desired materials. The sensor response was noted as a difference in resistance when the sensor

(a)

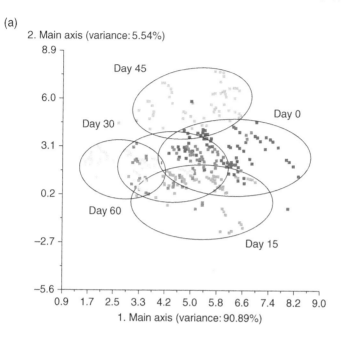

(b)

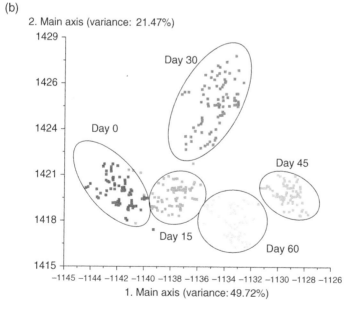

Figure 8.5 (a) PCA and (b) LDA analysis for Mandarins (80 samples) Adapted from [79].

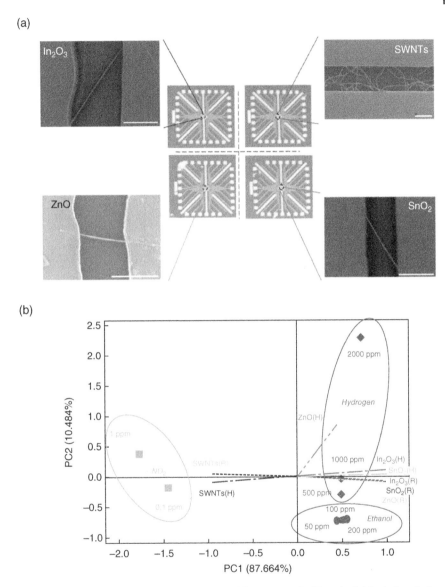

Figure 8.6 (a) Hybrid chemical sensor array chip composed of four individual chemical sensors, including individual In$_2$O$_3$ nanowire, SnO$_2$ nanowire, ZnO nanowire, and SWNT chemical sensor chips. The source–drain electrode distance is about 3 μm and (b) PCA scores and loading plots of the chemical sensor array composed of four different nanostructure materials and only three metal oxide nanowires Adapted from [48].

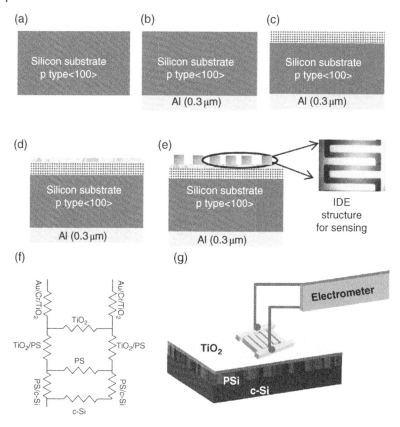

Figure 8.7 (a) Wafer cleaning. (b) Back side Al deposition for PS fabrication. (c) PS fabrication. (d) TiO$_2$ deposition on PS. (e) Au/Cr for electrode fabrication. (f) Electrical resistance equivalent model. (g) 3D schematic of TiO$_2$/PS sensor Adapted from [55].

was exposed to different organic vapors within the concentration range of 5–500 ppm in simulated real breath conditions. The sensing equation used was

$$S(\%) = \frac{R_a - R_0}{R_0} \times 100\% \tag{8.1}$$

where R_a and R_0 are the resistance change in presence and absence of analyte, respectively.

The results were recorded at room temperature, however, the tests were done at different temperatures and the optimized operating temperature was found to be around 100 °C. The sensors were exposed to vapors of ethanol, acetone, IPA, xylene, and benzene. The vapour pressures of different analytes were tuned for controlling their respective concentrations in ppm [40].

The TiO$_2$/PS sensor was tested at varied temperatures and different concentrations of ethanol. This showed that the maximum response was achieved at approximately $100\,°C$. The maximum response was shown to ethanol by TiO_2/PS with its response as 14% at 100 ppm at ethanol concentration. PS had shown higher response towards acetone in comparison to other tested VOCs. The sensors were exposed to different concentrations of ethanol vapors, for comparing all the sensors and checking their detection limit. The sensitivity of TiO_2/PS was found to be ~8.5 unit/ppm. Multiple studies done for repeated cycles and for almost six months fetched repetitive and stable results (Figure 8.8a). Different devices were selected from the same wafer to confirm the uniformity of the fabrication process and few devices were picked up from different wafers (Figure 8.8b). These sensing tests have also imparted good uniformity and reproducibility of the process.

The obtained sensing data from all three samples were combined to estimate the concentration of the vapors using polynomial regression model. The following equation was used:

$$\text{ppm}_{\text{gas}} = a_1 x + a_2 x^2 + \cdots + a_n x^n \tag{8.2}$$

Here, a_is are the coefficients and x is the sensor output. These coefficients are obtained using least-squares approximation [80]. The fourth-order and third-order polynomial regression models fitted best to estimate ppm of ethanol and acetone vapors, respectively (Figure 8.8c). In order the make an e-nose system, PCA was done to reduce the dimensions in the form of cluster. PCA clearly discriminated ethanol vapors from a group of analyte vapors (Figure 8.8d). Also IPA, xylene, and benzene were also separated with clear demarcation boundaries. The inset of Figure 8.8d illustrates the completed fabricated sensor bonded on an IC header for connecting it to the interfacing and peripheral circuits.

A very recent experiment in the area of e-nose has been reported for detection of wine properties by using MOX-based sensors array and machine learning technique [32]. The classification of wines was done on the basis of areas of production, vintage years, fermentation processes, and varietals. Six different MOX sensors whose conductivity was used as a tool for sensing were used. Four machine learning algorithms like XGBoost, RF, SVM, and BPNN were used for classification and separation of samples. It was found that BPNN showed the best performances in distinguishing production areas and varietals and SVM was found to showcase the best performances in detecting vintages and fermentation processes.

Alizadeh and Zeynali [81] developed an e-nose which consisted of an array of polymer-based surface acoustic wave (SAW) sensors and used PCA, ANN, and PCA–ANN techniques for classification of chemical warfare agents. SAW-based sensors with 100 MHz delay line devices were developed. The polymers chosen were hexafluoro-2-propanol-substituted polysiloxane (SXFA), poly(epichlorohydrin) (PECH), and phenyl methyl polysiloxane (75% phenyl and 25% methyl) OV_{25}. The warfare agents classified were dimethyl methyl phosphonate (DMMP) and 1,5-dichloropentane as commonly considered simulants of nerve agents (GD, GB, VX) and mustard gas (HD). DMMP, DCP,

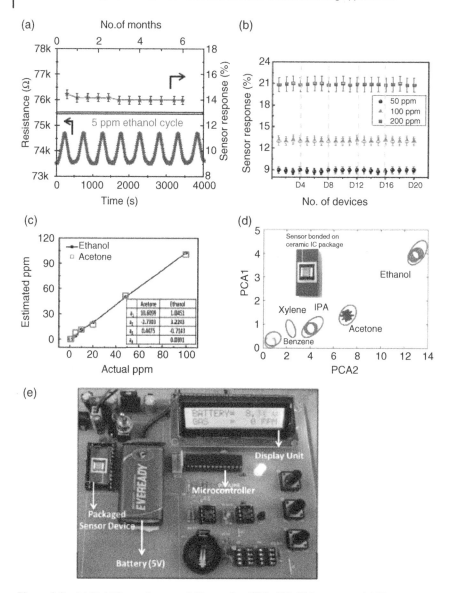

Figure 8.8 (a) Stability and repeatability study of TiO$_2$/PS. (b) Process scalability test. (c) Estimated and actual ppm for ethanol and acetone. (d) 2D PCA map showing ethanol vapor discrimination by the sensor, inset shows packaged sensor. (e) Lab prototype of TiO$_2$/PS sensor Adapted from [55].

and a binary mixture of these simulants were identified with 95% efficiency using optimized ANN and PCA–ANN techniques.

Gas sensor arrays have also been used to identify different aromas of grapes, apple, and strawberry due to their use in food applications [83]. Polyaniline films were used as sensitive layers and printed graphite IDEs (PGIEs) were used for higher sensitivity. The sensing response was also taken from gold-IDE based devices; however, the PGIE-based devices were much more sensitive and showed a better limit of detection (LOD) than IDEs based. PCA was used for classification and different aromas were successfully distinguished. A total of 98.77% information was collected using the first two main components (PC1 + PC2). The highest amount of information (93.03%) was contributed by PC1 and thus the aroma and concentration analysis was based on this PC1. The VOCs from apple, grapes, and strawberries were distinctively discriminated.

8.6 Improvements and Challenges

Although e-nose is one of the finest solution as an error-free and good learning model, there are certain challenges and scope of improvement in this area. Most of the e-nose products are based on MOX sensors which have selectivity issues. Also, these sensors are to be heated at a certain temperature to make these operational. This increases the power requirement, fabrication complexity, and heat dissipation issues to the peripheral devices. Thus, the first challenge is to make room temperature-operated MOX sensors. This is now becoming possible with the evolution of nanomaterials and heterostructures. Also, e-nose uses sensor array and more the sensors in the array, more the data and learning of the system. Unfortunately, this also crops up the problem of space occupied by each sensor. Thus, miniaturization is the key to the issues of space constraint. The sensors are required to be small enough so that many sensors can be accommodated in a small area. One of the other challenges with e-nose, is discrimination and identification of analytes in small concentrations. The probability of system in identifying this vapor as another is very high due to interference of another vapor and intersection with each other in the measured space.

8.7 Conclusion

The chapter elucidates the history, explanation, examples, challenges, and future scope of the e-nose system. It is like a human nose in an electronic form. The demand of e-nose has been increasing rapidly over time. The field where an application of such a system is most relevant is aroma and odor identification and discrimination. This is mostly used in food industry, healthcare, and environment monitoring. The key advantages of e-nose systems are that they: (i) are noninvasive, (ii) are intelligent devices as they can take decisions, (iii) can discriminate among analytes, (iv) are learning (data) dependent and (v) accurate. The chapter explicates the application areas of the e-nose structure. The components of the e-nose system were discussed in detail like sensor array, signal processing models, and decision making techniques. The type of sensors and their challenges were also briefly mentioned in this section. Few recent and popular examples of the e-nose system used in chemical sensing were also cited. It can be concluded that e-nose has been in existence for a long time now; however, its use in urban and rural areas is still a challenge. Also, many times, use of e-nose is a trade-off between quality and price. Few e-nose systems can be expensive and so its use is linked to its importance for an application. E-noses are known to be very sensitive, smart, and intelligent devices; however, reduction in its cost and construction complexity has a great scope in the research and development field.

References

1 https://phys.org/news/2018-05-electronic-nose-variety-scents.html.
2 https://spectrum.ieee.org/the-human-os/biomedical/devices/meet-the-enose-that-actually-sniffs.
3 Gardner, J.W. and Bartlett, P.N. (1994). A brief history of electronic noses. *Sensors and Actuators B* 18–19: 211–220.
4 Wilson, A.D. and Baietto, M. (2009). Applications and advances in electronic-nose technologies. *Sensors* 9: 5099–5148.
5 Canhoto, O. and Magan, N. (2003). Potential for the detection of microorganisms and heavy metals in potable water using electronic nose technology. *Biosensors & Bioelectronics* 18: 751–754.
6 Casalinuovo, I.A., Di'Piero, D., Coletta, M., and Di'Francesco, P. (2006). Application of electronic noses for disease diagnosis and food spoilage detection. *Sensors* 6: 1428–1439.
7 Paolesse, R., Alimelli, A., Martinelli, E. et al. (2006). Detection of fungal contamination of cereal grain samples by an electronic nose. *Sensors and Actuators B* 119: 425–430.
8 Turner, A.P.F. and Magan, N. (2004). Electronic noses and disease diagnostic. *Nature Reviews. Microbiology* 2: 161–166.
9 https://www.britannica.com/science/chemoreception/Smell.

10 Gardner, J.W. and Taylor, J.E. (2009). Novel convolution-based signal processing techniques for an artificial olfactory mucosa. *IEEE Sensors* 9: 929–935.

11 Dietrich, A.M. (2009). The sense of smell: contributions of orthonasal and retronasal perception applied to metallic flavor of drinking water. *Journal of Water Supply: Research and Technology* 58: 562–570.

12 Persaud, K. and Dodd, G.H. (1982). Analysis of discrimination mechanisms of the mammalian olfactory system using a model nose. *Nature* 299: 352–355.

13 Ikegami, A. and Kaneyasu, M. (1985). Olfactory detection using integrated sensors. *Proceedings of the 3rd International Conference on Solid-State Sensors and Actuators (Tranducers 1985), Philadelphia, PA, USA* (7–11 June 1985), pp. 136–139.

14 Deswal, A., Deora, N.S., and Mishra, H.N. (2014). Electronic nose based on metal oxide semiconductor sensors as an alternative technique for the spoilage classification of oat milk. *World Academy of Science, Engineering and Technology, International Journal of Agricultural and Biosystems Engineering* 8: 506–509.

15 Dhanekar, S., Sharma, I., and Islam, S.S. (2016). Optical measurement of trace level water vapours using functionalized porous silicon: selectivity studies. *RSC Advances* 6: 72371–72377.

16 Dhanekar, S., Islam, S.S., and Harsh (2012). Photo-induced electrochemical anodization of p-type silicon: achievement and demonstration of long term surface stability. *Nanotechnology* 23: 235501. (8 pp.).

17 Zhao, S., Li, Z., Wang, G. et al. (2018). Highly enhanced response of MoS_2/porous silicon nanowire heterojunctions to NO_2 at room temperature. *RSC Advances* 8: 11070–11077.

18 Zhang, T., Liu, G.-Q., Leong, W.-H. et al. (2018). Hybrid nanodiamond quantum sensors enabled by volume phase transitions of hydrogels. *Nature Communications* 9: 1–8.

19 Yin, J., Tan, Z., Hong, H. et al. (2018). Ultrafast and highly sensitive infrared photodetectors based on two-dimensional oxyselenide crystals. *Nature Communications* 9: 1–7.

20 Eranna, G. (2016). *Metal Oxide Nanostructures as Gas Sensing Devices*. Boca Raton: CRC Press.

21 Tian, W.-C., Ho, Y.-H., Chen, C.-H., and Kuo, C.-Y. (2013). Sensing performance of precisely ordered TiO_2 nanowire gas sensors fabricated by electron-beam lithography. *Sensors* 13: 865–874.

22 De Marcellis, A., Ferri, G., Member, S. et al. (2013). WO_3 hydrogen resistive gas sensor and its wide-range current-mode electronic read-out circuit. *IEEE Sensors Journal* 13: 2792–2798.

23 Li, C., Lv, M., Zuo, J., and Huang, X. (2015). SnO_2 highly sensitive CO gas sensor based on quasi-molecular-imprinting mechanism design. *Sensors* 15: 3789–3800.

24 Zappa, D., Galstyan, V., Kaur, N. et al. (2018). Metal oxide -based heterostructures for gas sensors – a review. *Analytica Chimica Acta* 1039: 1–23.

25 Dwivedi, P., Dhanekar, S., Das, S., and Chandra, S. (2017). Effect of TiO_2 functionalization on nano-porous silicon for selective alcohol sensing at room temperature. *Journal of Materials Science and Technology* 33: 516–522.

26 Dwivedi, P., Dhanekar, S., and Das, S. (2018). MoO_3/nano-Si heterostructure based highly sensitive and acetone selective sensor prototype: a key to non-invasive detection of diabetes. *Nanotechnology* 29: 275503. (9 pp.).

27 Dwivedi, P., Arya, D.S., Dhanekar, S., and Das, S. (2018). Development of scalable planar MEMS technology for low power operated selective ethanol sensor. *Journal of Micromechanics and Microengineering* 28: 105020. (10 pp.).

28 Schaller, E., Bosset, J.O., and Esher, F. (1998). "Electronic noses" and their application to food. *LWT – Food Science and Technology* 31: 305–316.

29 https://towardsdatascience.com/support-vector-machine-introduction-to-machine-learning-algorithms-934a444fca47.

30 Thanasarn, T. and Warisarn, C. (2013). Comparative analysis between BP and LVQ neural networks for the classification of fly height failure patterns in HDD manufacturing process. *ICEAST 2013,* Bangkok, Thailand.

31 Li, C.W. and Wang, G.D. (2006). The research on artificial olfaction system electronic nose. *Journal of Physics: Conference Series* 48: 667–670.

32 Liu, H., Li, Q., Yan, B. et al. (2019). Bionic electronic nose based on MOS sensors array and machine learning algorithms used for wine properties detection. *Sensors* 19: 45. (pp. 1–11).

33 https://dataaspirant.com/2017/05/22/random-forest-algorithm-machine-learing/

34 Kpwaiski, B.R. and Bender, C.F. (1972). Pattern recognition: a powerful approach to interpreting chemical data. *Journal of the American Chemical Society* 94: 5632–5639.

35 Abe, H., Kanaya, S., Takahashi, Y., and Sasaki, S.I. (1988). Extended studies of the automated odour-sensing system based on plural semiconductor gas sensors with computerized pattern recognition techniques. *Analytica Chimica Acta* 215: 155–168.

36 Gardner, J.W. (1991). Detection of vapours and odours from a multisensor array using pattern recognition: principal component and cluster analysis. *Sensors and Actuators* 4: 109–115.

37 Ouellette, J. (1999). Electronic noses sniff our new markets. *The Industrial Physicist* 5: 26–29.

38 Yea, B., Konishi, R., Osaki, T., and Sugahara, K. (1994). The discrimination of many kinds of odor species using fuzzy reasoning and neural networks. *Sensors and Actuators B* 45: 159–165.

39 Dhanekar, S., Islam, S.S., Julien, C.M., and Mauger, A. (2016). Morphology dependent PL quenching of multi-zone nanoporous silicon: size variant silicon nanocrystallites on a single chip. *Materials and Design* 101: 152–159.

40 Dwivedi, P., Dhanekar, S., and Das, S. (2016). Synthesis of α-MoO$_3$ nano-flakes by dry oxidation of RF sputtered Mo thin films and their application in gas sensing. *Semiconductor Science and Technology* 31: 115010. (7 pp.).

41 Freund, M.S. and Lewis, N.S. (1995). A chemically diverse conducting polymer-based electronic nose. *Proceedings of the National Academy of Sciences* 92: 2652–2656.

42 Hatfield, J.V., Neaves, P., Hicks, P.J. et al. (1994). Towards an integrated electronic nose using conducting polymer sensors. *Sensors and Actuators B* 18: 221–228.

43 Yim, H.S., Kibbey, C.E., Ma, S.C. et al. (1993). Polymer membrane-based ion-, gas-, and bio-selective potentiometric sensors. *Biosensors & Bioelectronics* 8: 1–38.

44 Gliszczyńska-Świgło, A. and Chmielewsk, J. (2017). Electronic nose as a tool for monitoring the authenticity of food: a review. *Food Analytical Methods* 10: 1800–1816.

45 Wojnowski, W., Majchrza, T., Dymerski, T. et al. (2017). Portable electronic nose based on electrochemical sensors for food quality assessment. *Sensors (Basel)* 17: 2715. (pp. 1–14).

46 Pavlou, A.K. and Turner, A.P. (2000). Sniffing out the truth: clinical diagnosis using the electronic nose. *Clinical Chemistry and Laboratory Medicine* 38: 99–112.

47 Kurup, P.U. (2008). An electronic nose for detecting hazardous chemicals and explosives. *IEEE Conference on Technologies for Homeland Security*, Waltham, MA, USA (12–13 May 2008).

48 Chen, P.-C., Ishikawa, F.N., Chang, H.-K. et al. (2009). Nano electronic nose: a hybrid nanowire/carbon nanotube sensor array with integrated micromachined hotplates for sensitive gas discrimination. *Nanotechnology* 20: 125503. (pp. 1-8).

49 Wilson, A.D. (2013). Diverse applications of electronic-nose technologies in agriculture and forestry. *Sensors* 13: 2295–2348.

50 Branca, A., Simonian, P., Ferrante, M. et al. (2003). Electronic nose based discrimination of a perfumery compound in a fragrance. *Sensors and Actuators B* 92: 222–227.

51 Wilson, A.D. (2012). Review of electronic-nose technologies and algorithms to detect hazardous chemicals in the environment. *Procedia Technology* 1: 453–463.

52 Wilson, A.D. and Baietto, M. (2011). Advances in electronic-nose technologies developed for biomedical applications. *Sensors* 11: 1105–1176.

53 De Vito, S., Massera, E., Quercia, L., and Di Francia, G. (2007). Analysis of volcanic gases by means of electronic nose. *Sensors and Actuators B* 127: 36–41.

54 Kızıl, Ü., Genç, L., and Aksu, S. (2017). Air quality mapping using an e-nose system in Northwestern Turkey. *Agronomy Research* 15: 205–218.

55 Dwivedi, P., Dhanekar, S., Agrawal, M., and Das, S. (2018). Interfacial engineering in TiO$_2$/nano-Si heterostructure based device prototype for e-nose application. *IEEE Transactions on Electron Devices* 65: 1127–1131.

56 Ohloff, G. (1990). *Smelling Materials and Sense of Smell*, 1–37. Berlin, Germany: Springer-Verlag.

57 Desor, J.A. and Beauchamp, G.K. (1974). The human capacity to transmit olfactory information. *Perception & Psychophysics* 16: 551–556.

58 Wang, X.R., Lizier, J.T., Berna, A.Z. et al. (2015). Human breath-print identification by e-nose, using information-theoretic feature selection prior to classification. *Sensors and Actuators B* 217: 165–174.

59 Kou, L., Zhang, D., and Liu, D. (2017). A novel medical e-nose signal analysis system. *Sensors* 17: 402. (pp. 1–15).

60 http://www.enose.nl/products/aeonose.

61 van de Goor, R.M.G.E., Leunis, N., van Hooren, M.R.A. et al. (2017). Feasibility of electronic nose technology for discriminating between head and neck, bladder, and colon carcinomas. *European Archives of Oto-Rhino-Laryngology* 274: 1053–1060.

62 https://airsense.com/en/products/portable-electronic-nose.

63 Spinelle, L., Gerboles, M., Kok, G. et al. (2017). Review of portable and low-cost sensors for the ambient air monitoring of benzene and other volatile organic compounds. *Sensors* 17: 1520. (pp. 1-30).

64 https://airsense.com/en/products/olfosense.

65 https://www.alpha-mos.com/new-heracles-neo-electronic-nose.

66 http://www.sensigent.com/products/cyranose.html.

67 Dutta, R., Hines, E.L., Gardner, J.W., and Boilot, P. (2002). Bacteria classification using Cyranose 320 electronic nose. *Biomedical Engineering Online* 1: 1–7.

68 https://www.estcal.com.

69 Baldwin, E.A., Bai, J., Plotto, A., and Dea, S. (2011). Electronic noses and tongues: applications for the food and pharmaceutical industries. *Sensors* 11: 4744–4766.

70 https://www.owlstoneinc.com/products/lonestar.

71 Pennazza, G. and Santonico, M. (eds.) (2019). *Breath Analysis*. San Diego, CA: Elsevier/Academic Press.

72 https://www.comon-invent.com.

73 http://sensorsgroup.uniroma2.it/artificial-olfaction-systems.

74 http://www.gsg-analytical.com/english/moses2.htm.

75 https://www.draeger.com/es_es/Applications/Products/Mobile-Gas-Detection/Multi-Gas-Detection-Devices/X-am7000.

76 http://www.raesystems.com/products/multirae.

77 Sanchez-Garrido, C., Monroy, J.G., and Gonzalez-Jimenez, J. (2014). A configurable smart e-nose for spatio-temporal olfactory analysis. In: *2014 IEEE Sensors, Valencia, Spain (2–5 November 2014)*, 1968–1971. IEEE.

78 Wojnowski, W., Majchrzak, T., Dymerski, T. et al. (2017). Portable electronic nose based on electrochemical sensors for food quality assessment. *Sensors* 17: 2715. (pp. 1-14).

79 Gómez, A.H., Wang, J., Hu, G., and Pereira, A.G. (2006). Electronic nose technique potential monitoring mandarin maturity. *Sensors and Actuators B* 113: 347–353.

80 Chen, P.-C., Shen, G., and Zhou, C. (2006). Chemical sensors and electronic noses based on 1-D metal oxide nanostructures. *IEEE Transactions on Nanotechnology* 7: 668–682.

81 Manolakis, D.G., Ingle, K.V., and Kogon, M.S. (2000). *Statistical and Adaptive Signal Processing: Spectral Estimation, Signal Modeling, Adaptive Filtering and Array Processing*, 1e, 97–105. Boston, MA: Arctech House.

82 Alizadeh, T. and Zeynali, S. (2008). Electronic nose based on the polymer coated SAW sensors array for the warfare agent simulants classification. *Sensors and Actuators B* 129: 412–423.

83 Graboski, A.M., Ballen, S.C., Galvagni, E. et al. (2019). Aroma detection using a gas sensor array with different polyaniline films. *Analytical Methods* 11: 654–660.

9

Odor Sensing System

Takamichi Nakamoto and Muis Muthadi

Institute of Innovative Research, Tokyo Institute of Technology, Japan

9.1 Introduction

Although we have good sensors related to vision and audition, sensors related to olfaction is still under progress. However, an odor sensing system is required in many fields such as food, beverage, fragrance, health care, and environmental testing. Although many attempts to realize an electronic nose have been performed for three decades [1], sensors with higher capabilities are still required.

Persaud proposed the method to discriminate among odors using plural sensors and pattern recognition [2]. Then, our group proposed the method using plural sensors and neural network pattern recognition [3]. Thereafter, many researchers entered the field of electronic nose and the community of electronic nose has been formed.

A variety of sensors have been studied to detect and identify odorants. They are metal oxide gas sensors [4], conducting polymer sensors [5], metal oxide semiconductor (MOS) field effect transistor (FET) gas sensors [6], electrochemical sensors [7], quartz crystal microbalance (QCM) sensors [8], surface acoustic wave (SAW) sensors [9], cantilever-type sensors [10], near infrared spectroscopy [11], and optical sensors utilizing fluorescence, absorption [12–14].

However, there is still room for improvement in terms of sensitivity, selectivity, and stability. One of the solutions is a biologically inspired approach. Thus, we have studied an odor biosensor composed of cells expressing olfactory receptors (ORs).

Moreover, odor impression can be predicted from the sensing information. Mapping of sensing information such as mass spectrum onto odor impression in the sensory space using deep learning is described.

Smart Sensors for Environmental and Medical Applications, First Edition. Edited by Hamida Hallil and Hadi Heidari.

Then, another task of odor sensing to localize an odor source is explained. The odor source localization is a difficult task since turbulent air flow bothers the exploratory behavior of the odor source. The strategy of odor source localization has been developed on the platform of computational fluid dynamics (CFD).

In this chapter, recent hot topics such as odor biosensor, prediction of odor impression, and strategy for odor-source localization are described.

9.2 Odor Biosensor

An OR is typically utilized in an odor biosensor [15, 16]. An OR does not have high selectivity of odorants, whereas biological materials such as enzymes and antibodies have very narrow specificities. A single OR does not have enough specificity to discriminate among odorants.

The principle of odor recognition using ORs is shown in Figure 9.1. It is thought that ORs can recognize a part of the stereo chemical structure of the odorant. Although each OR responds to multiple odorants, the response pattern from the multiple ORs is distinguishable. Its specificity is broader than a conventional biosensor based upon enzyme-substrate reaction or antigen–antibody reactions. However, the selectivity of each OR is still higher than that of the conventional artificial sensor. The output pattern of several ORs is recognized using a pattern recognition algorithm.

There are three types of odor biosensors such as tissue-based sensor, receptor-based sensor, and cell-based sensor. In tissue-based sensor, olfactory epithelium and mucosa cut from the biological organ is mounted on a transducer [17].

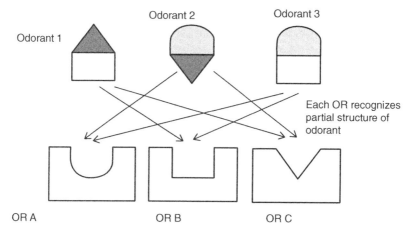

Figure 9.1 Principle of an odor biosensor using pattern recognition.

Extracellular potential of the olfactory epithelium and mucosa is detected by a microelectrode array as is shown in Figure 9.2a. The receptor-based sensor utilizes an OR as is shown in Figure 9.2b. One of the examples of transducers is a carbon nanotube transistor [18]. The cell-based sensor has an OR embedded in a cell. The cell can be mounted on a transducer such as QCM, SPR, and FET as is shown in Figure 9.2c.

In this chapter, the cell-based odor biosensor is described. Its principle is shown in Figure 9.3. This is based upon an insect OR since its mechanism is simpler than that of a mammal. The cell used here was Sf21 and the OR was of a Drosophila. When an odorant is captured by an OR, the ion channel is open and the flux of

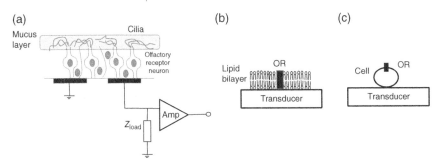

Figure 9.2 Types of odor biosensors: (a) tissue-based, (b) receptor-based, and (c) cell-based odor biosensors.

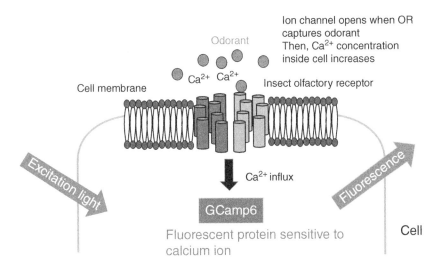

Figure 9.3 Principle of cell-based odor biosensor using fluorescence.

(a)   (b)

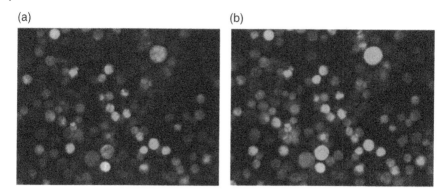

Figure 9.4   Image of cells expressing OR (a) before odorant exposure and (b) after exposure.

calcium ion goes into the cell. Since there is a fluorescent protein sensitive to calcium ion inside the cell, the fluorescent light (515 nm, green light) is generated when the excitation light (488 nm, blue light) is illuminated [19]. The fluorescent light is captured by C-MOS camera through a dichroic mirror. The detail of measurement system is described in Refs. [20, 21].

Figure 9.4a shows the fluorescent image of Sf21 cells expressing both BmOR3 and OR13a before odorant exposure. BmOR3 responds to bombykal and OR13a responds to 1-octen-3-ol, respectively. The cells expressing different ORs are randomly distributed. Figure 9.4b shows the cell image after 100 μM 1-octen-3-ol exposure. The brightness of the cell expressing OR13a increased after odorant exposure. The odorants can be classified using the image recognition technique. Moreover, the lock-in measurement technique enhanced the signal-to-noise ratio [21].

9.3   Prediction of Odor Impression Using Deep Learning

Although the main task of an odor sensing system is odor classification, the new topic in this field is the prediction of odor impression. The data in sensing space is mapped onto sensory space using this technique as is shown in Figure 9.5. Since an odor biosensor described in the previous section is at the fundamental stage, we used a mass spectrometer to obtain the data in sensing space. Although molecular structure parameters were used by other groups [22, 23], we use mass spectrometry since we can easily handle mixtures. It is possible to obtain large-scale data using mass spectrometry since its measurement method has been already established. The data in sensory space is obtained from the sensory test. Here, we used the data collected by Dravnieks [24].

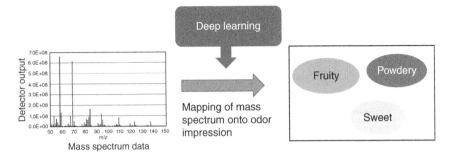

Figure 9.5 Principle of odor-impression prediction.

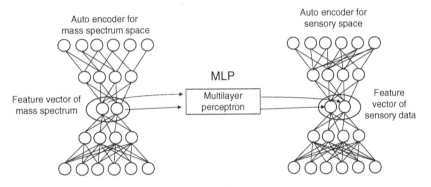

Figure 9.6 Prediction model of odor impression using two auto encoders.

The mapping from sensing space onto sensory space was performed using deep learning technique. Since the mapping mentioned above is highly nonlinear, the linear technique such as multiple linear regression is insufficient [25]. Although we previously used a neural network to predict the sensory quantity [26], it was limited to small-scale data. There are many studies related to deep learning [27, 28]. However, the original deep learning technique using two auto encoders was developed for our purpose [29]. The structure of prediction model is shown in Figure 9.6. One of the auto encoders is used to extract the feature of sensing space and the other extracts the feature of sensory space. Both features appear in the middle hidden layers and the former feature is mapped onto the latter one using a multilayer perceptron (MLP). The noise can be reduced in both spaces since their features are extracted. Both auto encoders have five layers and the MLP for mapping has five layers. Totally a nine-layer MLP is used to predict odor impression.

Its mechanism is inspired by partial linear regression (PLS) [30], which is a gold standard in chemometrics. In PLS, the features in independent variable space and the features in dependent variable space are used. The latter ones are obtained from the former ones using multiple linear regression. These features are called latent variables.

It is known that the prediction accuracy of PLS is superior to that of multiple linear regression [31]. Although our model has a structure similar to PLS, its prediction accuracy is expected to be further improved due to its nonlinearity.

Next, the experiment on predicting odor impression was performed. Mass spectra of 121 chemicals were obtained from the Chemistry WebBook provided by National Institute of Standards and Technology [32]. The ionization method was EI method with its energy of 70 eV. Since intensities with mass to charge ratio from 51 to 262 were used, the data in sensing space has 212 dimensions. Each chemical has scores of 144 descriptors described in Ref. [24].

The structure of the neural network was optimized. To give the predictive model a sufficient margin, the features in sensory space has 30 dimensions and the feature in mass spectrum data space has 45 dimensions. The evaluation was performed using sixfold cross validation. Figure 9.7a shows the prediction accuracy of our proposed model, whereas Figure 9.7b shows the prediction accuracy using PLS. The number of latent variables in PLS was 45 after the optimization. The predicted values were correlated with human sensory evaluation with $R = 0.76$ in the proposed model, while the correlation coefficient for PLS model was about 0.61. Thus, the prediction accuracy of our proposed model was better than that of conventional techniques such as PLS.

The prediction accuracy seems to be improved if more data are used for training. However, many of the data related to odor impression have only odor descriptors and the degree of odor impression is not included. Thus, odor descriptors similar to each other appear mutually exclusive and the prediction of odor impression is difficult even if large-scale data are available. Thus, we made odor-descriptor groups composed of similar odor descriptors and the prediction of odor-descriptor group was performed.

The input data used in this experiment were the mass spectrum dataset used in this study, which was again obtained from the Chemistry Webbook provided by National Institute of Standards and Technology. Odor descriptors corresponding to a mass spectrum were obtained from "Flavors and Fragrances" catalog published from Sigma-Aldrich [33]. The descriptor is given in a binary form in the Sigma-Aldrich catalog, whereas it is given on a scale of 0–5 in Dravnieks' sensory evaluation used in the previous study. We had the data of 999 samples and the number of descriptors used here was 138.

Then, we tried to predict the group of odor descriptors closely similar to each other. Thus, we should know the similarity between odor descriptors. The similarity was calculated by the natural language modeling method called word2vec. Word2vec is a language modeling method proposed by Mikolov et al. at Google Inc. [34]. The model uses a three-layer neural network called a skip-gram model

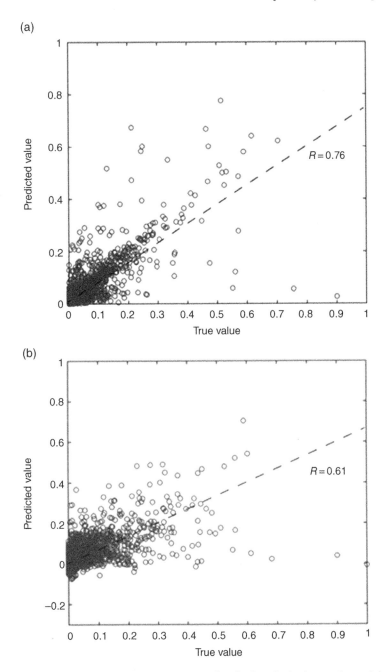

Figure 9.7 Relationship between the predicted value of odor impression and the actual score after normalization. (a) Proposed model with two auto encoders and (b) PLS model [29].

trained with a large text corpus and is used to obtain a vector representation of words. In the Word2vec training scheme, the proximity between odor descriptors in the word vector space is high if many words share a common context in the training corpus. The full text of English Wikipedia (enwiki-201509001, 12.4 GB [35]) corpus was used here as a training corpus.

We used two matrices to express similarities between all possible pairs of odor descriptors [36]. One was the matrix with correlation coefficients and the other was the matrix with cosine distance between vectors obtained from Word2vec. Then, hierarchical clustering was performed to obtain odor-descriptor groups. We can change the number of odor descriptor groups when we modify the cutoff point.

We used a two-step neural network model. The feature of mass spectrum was extracted using the auto encoder. The feature dimension was 30 after the optimization. Then its feature was mapped onto the odor descriptor group using a four-layer MLP. The evaluation of the predictive model was performed using a fivefold cross-validation technique.

Figure 9.8 shows the accuracy of an odor descriptor group prediction as a function of number of clusters. True positive means the rate where the model outputs 1 when the desired output is 1, and true negative means the rate where the model

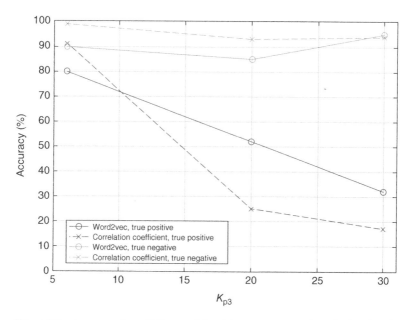

Figure 9.8 Accuracy of predictive models with respect to number of odor descriptor groups when we make a group using natural language processing [36].

outputs 0 when the desired output is 0. Since more than 98.4% of the sensory dataset have zeros, the predictive models tend to output 0. Thus, the accuracy of true positives tends to become lower.

The accuracies of the two predictive models based upon correlation coefficients and Word2vec are compared in Figure 9.8. Although the number of clusters has a trade-off relationship with the accuracy of the model, the prediction accuracy of true positives is 53% and the prediction accuracy of true negatives is 85% using natural language processing when the number of clusters is 20, while the model with correlation-coefficient-based clusters shows lower accuracy. Thus, the method using natural language processing is useful for predicting odor impressions.

9.4 Establishment of Odor-Source Localization Strategy Using Computational Fluid Dynamics

9.4.1 Background of Odor-Source Localization

The topic of robotic chemical sensing system was first brought up in the early 1980s by Larcombe and Halsall [37]. They reviewed the potential application of remotely controlled robot devices with a chemical sensing ability in hazardous environments, such as fuel processing, reactor maintenance, transportation of active material, waste handling, and incident management. Since then, interest on this research topic has grown and the study on robotic odor-source localization system has nowadays become an active research field.

An odor-source localization system becomes one of a highly demanded technology nowadays owing to its potential application in various services from safety operation, such as finding fire at the initial stage and dangerous odor leakage in a factory, to humanitarian tasks, such as finding human victims trapped under building rubbles. This system is expected especially for dangerous tasks; thus, it is not preferable to let a human or a tracking dog carry out the task.

Various strategies of odor-source localization are inspired by strategies of biological organisms which are also known as biologically inspired strategies. These organisms have different localization strategies best suited to their unique morphology, sensing features, and natural environmental conditions. Unicellular organisms have different strategies from the strategies of complex organisms. Some robotic strategies are developed by harnessing hardware and computational techniques that are not available in the biological organism [38, 39].

With so many variations of localization strategy, in order to make quantitative comparisons, it is important to test the strategies in a consistent environment. A consistent environment cannot be realized in the real environment. With the advancement in computation capability, it is now feasible to simulate a realistic virtual environment using CFD software. The odor-source localization strategy can be initially tested in the virtual environment before an actual test in the real environment. This testing method allows faster development-testing cycle. CFD has been widely used to simulate indoor and outdoor environment, allowing reasonably accurate assessment on environmental conditions such as odor distribution, airflow pattern, and heat transfer [40, 41].

The odor source localization is generally carried out in three stages:

1) Searching for odor information (trail or plume).
2) Tracking odor information toward the odor source.
3) Declaring the discovery of the odor source, which is generally called odor source declaration.

The localization typically experiences multiple odor detections and loses. Therefore, the act of searching for odor and tracking may need to be repeated multiple times before declaration of the source discovery.

Our work here is focused on searching for odor plume and tracking it toward the source. Declaring the discovery of the odor source based on sensory information is not an easy task as it is difficult to determine the true source from the false one. There have been many works reported to address this task. Therefore, a simple odor source declaration is used here, that is the odor source is simply declared when the sensor node arrives by a certain distance from the true source location. In the experiment described later, it was 20 cm.

9.4.2 Sensor Model with Response Delay

One of the shortcomings of the chemical sensor is the significant response delay. When the localization is carried out in the real world environment, in which odor distribution changes dynamically, the delay leads to inaccuracy in evaluating the real-time odor information. Therefore, this delay characteristic needs to be accommodated into the design of localization strategy.

A realistic sensor response model is required for the simulation. In this matter, Ishida et al. [42] derived a sensor model from a real experiment using a semiconductor gas sensor (TGS822, Figaro Engineering). The steady-state model of sensor response calibration over ethanol concentration is

$$r_0(k) = \left[1 + 0.2309 C_0(k)\right]^{-0.6705} \tag{9.1}$$

where $r_0(k)$ is the ideal sensor response (ratio of R_{gas} to R_{air}) at time k and C_0 is the ethanol concentration in ppm. The dynamic sensor response including the delay can be appropriately described using the first-order discrete time system

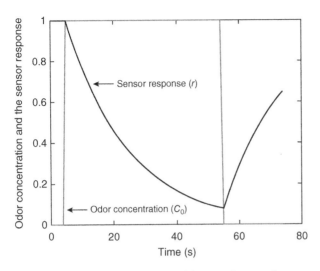

Figure 9.9 Response of sensor model to a unit step odor concentration signal.

$$r\big(k+1\big)=0.95\big(k\big)+0.05r_0\big(k\big) \tag{9.2}$$

where $r(k)$ and $r(k+1)$ are the observed sensor response at time k and $k+1$, respectively. Using Eqs. (9.1) and (9.2), the characteristic of sensor response $r(k)$ with a unit step of odor concentration $C_0(k)$ as the input can be shown in Figure 9.9. The time constant of Eq. (9.2) is 19.5 seconds.

9.4.3 Simulation of Testing Environment Using CFD

Indoor environments typically have a less complex airflow pattern than outdoor environment. However, this does not mean odor source localization in indoor environment is an easy task. Besides, there are a lot of tasks in which an indoor odor source localization system can be very helpful.

We have studied a localization strategy for indoor environment. Using a CFD software package, ANSYS Fluent 17.2, we created a virtual environment based on a closed room used in [43] with three obstacles as the testing environment as shown in Figure 9.10. In this closed room, an odor source is located on the floor between a window and three obstacles. The window has colder temperature, 10 °C, while the other parts of the room has a temperature of 15 °C. This temperature difference induced a circulating airflow vertically as a heat transfer phenomenon.

In environments with medium to strong airflow, odor is mostly distributed by turbulent airflow. There are various turbulence models available. Here, a standard

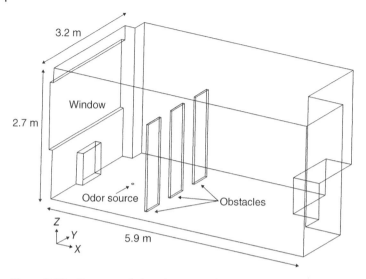

Figure 9.10 Geometry of closed room environment.

turbulent model, that is k-ε model based on Reynolds-Averaged Navier-Stokes (RANS) models was selected as it can represent the turbulence appropriately well with low computation load. This model is defined by two transport equations which describe the behavior of kinetic energy k and its dissipation rate ε [44]. Some other simulation parameters are listed in Table 9.1.

It is assumed that the room is initially without airflow. As the convective airflow developed gradually with time, in order to get an odor distribution in fairly developed convective airflow, the simulation in CFD was first performed without odor

Table 9.1 Simulation parameters.

Parameter	Setting
Odor source	Ethanol gas (C_2H_5OH) Emission rate: 50 ml/min
Material	Room boundaries: aluminum Fluid: mixture of ethanol gas and air
Temperature	Window: 10 °C Others: 15 °C
Turbulence model	Standard k-ε model
Computation time step	0.1 s

release from the source for 600 seconds long. In the end of this simulation, a relatively steady airflow was developed. The simulation was continued with odor release for another 600 seconds. In the end of the simulation, we have data of the simulated environment with odor release containing odor distribution and airflow information.

9.4.4 Simulation of Biologically Inspired Odor-Source Localization

After obtaining data from the CFD simulation, the process is then continued by localization simulation using MATLAB programming. To extend the data into more refined and uniform scaled array, the data is then processed with a grid interpolation in MATLAB. In this case the grid size is 5 cm, which is also the interval of odor measurement by the sensor node while moving during the localization task.

In this simulation, the sensor node's mobility is on the floor, with the odor sensor mounted 5 cm above the floor. Thus, a time-series 2D data at 5 cm above the floor was extracted from the 3D data. Moreover, only the data when odor had been released was used for the simulation. Figure 9.11 shows the 2D odor distribution and airflow profile at 100 seconds after the start of odor release.

9.4.4.1 Odor Plume Tracking Strategy

There are a wide range of variations in localization strategy by biological organisms. Among the most cited strategies is the pheromone localization strategy by flying silk moths. Silk moths' localization strategy comprises two key elements:

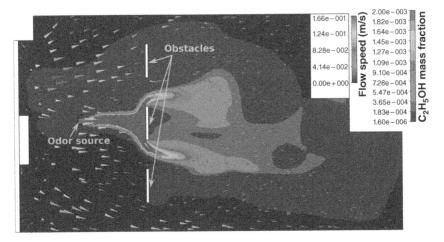

Figure 9.11 Odor distribution and airflow profile on a plane 5 cm above floor at 100 seconds after starting odor release.

moving in upwind direction when sensing pheromone trail and perform cross-wind search by moving in zigzag or circular pattern when losing the trail. Various reactive search strategies were inspired by the silk moths' strategy [45, 46].

Previously, we designed a reactive plume tracking strategy using a sensor without response delay [47]. However, when the implementation of this strategy was simulated for localization using a sensor with response delay in a turbulent environment, the performance was poor owing to the delayed odor detection. Thus, instead of using delayed odor detection to track the fluctuating plume directly, it is more reliable to use the detection to estimate the plume edges at the right and left sides and then approximate the plume centerline which is around the central point between the edges. We developed a plume edge estimation method based on odor gradient sign [48]. This method is robust against odor fluctuation and applicable for odor sensing with response delay. The plume tracking can be done by moving along the centerline toward upwind direction.

Taking inspiration from the moth's strategy and using our approach in tracking the plume, we designed an odor tracking strategy to localize the source in a closed room environment for robotic implementation using current sensing technology. The strategy comprises a series of three phases as described in Figure 9.12:

1) Upwind tracking: Upon the first detection, the sensor node moves in the upwind direction while measuring odor intensity at a regular interval of time or space. During this phase, if the sensor node fails to detect odor, the phase switches to "crosswind scanning."
2) Crosswind scanning: Upon beginning of crosswind scanning, the node moves in the crosswind direction back and forth to estimate the plume edges at the right and left sides. This back-and-forth movement is repeated with expanding length when the two edges has not been yet discovered. Upon finding the edges, the plume centerline can be approximated at the central point between the two edges and then the phase switches to "move to the centerline."
3) Move to the centerline: In this phase, the node simply moves to the approximated centerline. When the node has reached the estimated centerline, the phase switches to "upwind tracking."

This series of phases is repetitive until the odor source is found. A robot is assumed to have an ultrasonic sensor to avoid the collision with obstacle. The obstacle is avoided by turning 90° away from the direction to the obstacle. The robot moved at $5\,\mathrm{cm\,s}^{-1}$ during "upwind tracking" and "move to the centerline" and $2.5\,\mathrm{cm\,s}^{-1}$ during "crosswind scanning."

9.4.4.2 Result

The strategy was tested by performing 404 trials of localization started from different locations with a duration length of 500 seconds. The quantitative performance

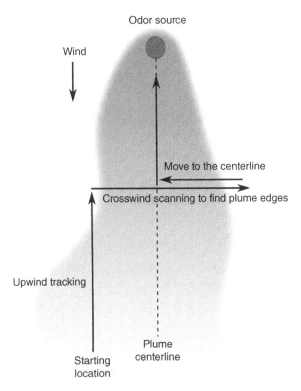

Figure 9.12 Phases of odor plume tracking.

was determined by the success rate as well as the time required to find the source (time performance). Figure 9.13 shows the performance of a localization strategy with different starting positions. The localization is successful if the odor source is found before the time limit of 500 seconds; otherwise, it fails. The lightest grayscale level in the figure signifies the failure. Most of the success cases are the localizations started from position around the plume centerline. The statistics showed that our localization strategy, which, considering response delay, achieved a success rate of 71.53% with average time of localization 269.02 seconds. It is a significant improvement compared to the strategy implementation without response delay consideration which achieved a success rate of 63.61% with average time of localization 331.80 seconds.

9.4.5 Summary of Odor Source Localization Strategy

Wide range of odor-source localization strategies are developed for different environmental conditions and different sensing technology. Evaluation of these strategies on a real environment is cumbersome and impractical. The advancement in

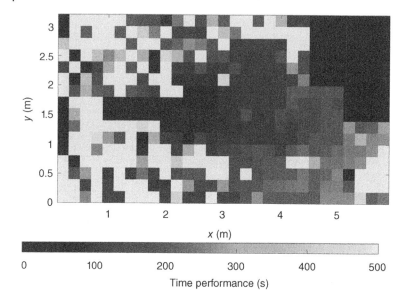

Figure 9.13 Performance distribution with variation of starting position when sensor response delay was considered. The color and position of cells represent the time performance of localization and the starting position, respectively. *Source:* Reprinted with permission from Ref. [47]. Copyright MYU TOKYO, 2018.

CFD simulation proposes a faster cycle of development-testing of odor-source localization strategy before an actual test on a real environment. This study has demonstrated the CFD application for development and testing the localization strategy for a small environment. However, the application for a larger and more complex environment should be aimed for further research.

9.5 Conclusion

In this chapter, recent hot topics such as odor biosensor, prediction of odor impression, and strategy for odor-source localization were described. Recently biologists contribute to odor biosensors. Computer scientists join the field to predict odor impression. Mechanical and electronic engineers publish many reports of odor-source localization. The researcher populations of those fields were gradually increasing in those interdisciplinary areas and the new field will be open when several different areas are merged.

Acknowledgments

The authors wish to thank Prof. Ryohei Kanzaki and Dr. Hidefumi Mitsuno for providing cells expressing ORs. This work was partially supported by JSPS KAKENHI (Grant Number JP18H03773).

References

1 Pearce, T.C., Schiffman, S.S., Nagle, H.T., and Gardner, J.W. (eds.) (2006). *Handbook of Machine Olfaction: Electronic Nose Technology*. Wiley-VCH.

2 Persaud, K. and Dodd, G. (1982). Analysis of discrimination mechanisms in the mammalian olfactory system using a model nose. *Nature* 299: 352–355.

3 Nakamoto, T. and Moriizumi, T. (1988). Odor sensor using quartz-resonator array and neural-network pattern recognition. *IEEE 1988 Ultrasonics Symposium Proceedings*, pp. 613–616.

4 Kaneyasu, M., Ikegami, A., Arima, H., and Iwanaga, S. (1987). Smell identification using a thick-film hybrid gas sensor. *IEEE Transactions on Components, Hybrids, and Manufacturing Technology* 10: 267–273.

5 Hatfield, J.V., Neaves, P., Hicks, P.J. et al. (1994). Towards an integrated electronic nose using conducting polymer sensors. *Sensors and Actuators B: Chemical* 18: 221–228.

6 Sundgren, H., Lundström, I., Winquist, F. et al. (1990). Evaluation of a multiple gas mixture with a simple MOSFET gas sensor array and pattern recognition. *Sensors and Actuators B: Chemical* 2: 115–123.

7 Stetter, J.R., Jurs, P.C., and Rose, S.L. (1986). Detection of hazardous gases and vapors: pattern recognition analysis of data from an electrochemical sensor array. *Analytical Chemistry* 58: 860–866.

8 Ema, K., Yokoyama, M., Nakamoto, T., and Moriizumi, T. (1989). Odour-sensing system using a quartz-resonator sensor array and neural-network pattern recognition. *Sensors and Actuators* 18: 291–296.

9 Grate, J.W. and Abraham, M.H. (1991). Solubility interactions and the design of chemically selective sorbent coatings for chemical sensors and arrays. *Sensors and Actuators B: Chemical* 3: 85–111.

10 Lange, D., Hagleitner, C., Hierlemann, A. et al. (2002). Complementary metal oxide semiconductor cantilever arrays on a single chip: mass-sensitive detection of volatile organic compounds. *Analytical Chemistry* 74: 3084–3095.

11 Buratti, S., Sinelli, N., Bertone, E. et al. (2015). Discrimination between washed Arabica, natural Arabica and Robusta coffees by using near infrared spectroscopy,

electronic nose and electronic tongue analysis. *Journal of the Science of Food and Agriculture* 95: 2192–2200.

12 Rakow, N.A. and Suslick, K.S. (2000). A colorimetric sensor array for odour visualization. *Nature* 406: 710–713.

13 Aernecke, M.J. and David, R. (2009). Walt, Optical-fiber arrays for vapor sensing. *Sensors and Actuators B* 142: 464–469.

14 Nakamoto, T., Yosihioka, M., Tanaka, Y. et al. (2006). Colorimetric method for odor discrimination using dye-coated plate and multiLED sensor. *Sensors and Actuators B* 116: 202–206.

15 Persaud, K.C. (2012). Biomimetic olfactory sensors. *IEEE Sensors Journal* 12: 3108–3112.

16 Park, T.H. (ed.) (2014). *Bioelectronic Nose*. Springer.

17 Liu, Q., Ye, W., Xiao, L. et al. (2010). Extracellular potentials recording in intact olfactory epithelium by microelectrode array for a bioelectronic nose. *Biosensors and Bioelectronics* 25: 2212–2217.

18 Goldsmith, B.R., Mitala, J.J. Jr., Josue, J. et al. (2011). Biomimetic chemical sensors using nanoelectronic readout of olfactory receptor proteins. *ACS Nano* 5: 5408–5416.

19 Mitsuno, H., Sakurai, T., Namiki, S. et al. (2015). Novel cell-based odorant sensor elements based on insect odorant receptors. *Biosensors & Bioelectronics* 65: 287–294.

20 Sukekawa, Y., Mujiono, T., Nakamoto, T. et al. (2016). Development of automated flow measurement system for cell-based odor sensor [in Japanese]. *IEEJ Transactions on Sensors and Micromachines* 136: 289–295.

21 Sukekawa, Y. and Nakamoto, T. (2019). Odor biosensor system based on image lock-in measurement for odorant discrimination. *Electronics and Communications in Japan* 102: 57–64.

22 Keller, A., Gerkin, R.C., Guan, Y. et al. (2017). Predicting human olfactory perception from chemical features of odor molecules. *Science* 355: 820–826.

23 Gutiérrez, E.D., Dhurandhar, A., Keller, A. et al. (2018). Predicting natural language descriptions of mono-molecular odorants. *Nature Communications* 9: 4979.

24 Dravnieks, A. (1985). *Atlas of Odor Character Profiles*. Philadelphia: ASTM.

25 Yokoyama, K. and Ebisawa, F. (1993). Detection and evaluation of fragrances by human reactions using a chemical sensor based on adsorbate detection. *Analytical Chemistry* 65: 673–677.

26 Hanaki, S., Nakamoto, T., and Moriizumi, T. (1996). Artificial odor-recognition system using neural network for estimating sensory quantities of blended fragrance. *Sensors and Actuators A* 57: 65–71.

27 Hinton, G.E. and Salakhutdinov, R.R. (2006). Reducing the dimensionality of data with neural networks. *Science* 313: 504–507.

28 Goodfellow, I., Bengio, Y., and Courville, A. (2016). *Deep Learning*. MIT Press.

29 Nozaki, Y. and Nakamoto, T. (2016). Odor impression prediction from mass spectra. *PLoS One* 11 (6): e0157030. https://doi.org/10.1371/journal.pone.0157030.

30 Geladi, P. and Kowalski, B.R. (1986). Partial least-squares regression: a tutorial. *Analytica Chimica Acta* 185: 1–17.

31 Carey, W.P., Beebe, K.R., Kowalski, B.R. et al. (1986). Selection of adsorbates for chemical sensor arrays by pattern recognition. *Analytical Chemistry* 58: 149–153.

32 "NIST Chemistry WebBook" [Online]. Available: http://webbook.nist.gov/chemistry/. [Accessed: 21-Dec-2019].

33 Sigma-Aldrich. *Flavors and Fragrances* [Online]. http://go.sigmaaldrich.com/ ff-catalogdownload-safcglobal.

34 Mikolov, T., Chen, K., Corrado, G., and Dean, J. (2013). Efficient estimation of word representations in vector space. ArXiv13013781 Cs (January 2013).

35 The Wikipedia Corpus [Online]. https://corpus.byu.edu/wiki., (Accessed: 27 December 2017)

36 Nozaki, Y. and Nakamoto, T. (2018). Predictive modeling for odor character of a chemical using machine learning combined with natural language processing. *PLoS One* 13 (6): e0198475. https://doi.org/10.1371/journal.pone.0198475.

37 Larcombe, M. and Halsall, J. (1984). *Robotics in Nuclear Engineering: Computer-assisted Teleoperation in Hazardous Environments with Particular Reference to Radiation Fields*. Graham & Trotman for the Commission of the European Communities, Directorate-General, Information Market and Innovation.

38 Li, J.-G., Meng, Q.-H., Wang, Y., and Zeng, M. (2011). Odor source localization using a mobile robot in outdoor airflow environments with a particle filter algorithm. *Autonomous Robots* 30 (3): 281–292.

39 Neumann, P., Bennetts, V.H., Lilienthal, A.J. et al. (2013). Gas source localization with a micro-drone using bio-inspired and particle filter-based algorithms. *Advanced Robotics* 27 (9): 725–738.

40 Chen, Q. and Srebric, J. (2000). Application of CFD tools for indoor and outdoor environment design [Invited paper]. *International Journal on Architectural Science* 1 (1): 14–29.

41 Li, X.F., Zhou, J., Shao, J., and Zhang, Q. (2010). CFD-based simulation analysis of indoor and outdoor ventilation environment. *International Conference on Computational Intelligence and Software Engineering.*, 10–12 December 2010, Wuhan, China.

42 Ishida, H., Nakamoto, T., and Moriizumi, T. (1998). Remote sensing of gas/odor source location and concentration distribution using mobile system. *Sensors and Actuators B: Chemical* 49 (1–2): 52–57.

43 Nakamoto, T., Ishida, H., and Matsukura, H. (2012). Olfactory display using solenoid valves and fluid dynamics simulation. In: *Multiple Sensorial Media Advances and Applications: New Developments in Mulsemedia*, 140–163. IGI Global.

44 Davidson, L. (2018). An Introduction to Turbulence Models. http://www.tfd.chalmers.se/~lada/postscript_files/kompendium_turb.pdf, p. 25. Chalmers (Publication 97/2). (accessed 21 December 2019).

45 Voges, N., Chaffiol, A., Lucas, P., and Martinez, D. (2014). Reactive searching and infotaxis in odor source localization. *PLoS Computational Biology* 10 (10): e1003861. https://doi.org/10.1371/journal.pcbi.1003861.

46 Russell, R.A., Bab-Hadiashar, A., Shepherd, R.L., and Wallace, G.G. (2003). A comparison of reactive robot chemotaxis algorithms. *Robotics and Autonomous Systems* 45: 83–97.

47 Muhtadi, M. and Nakamoto, T. (2017). Optimal estimation method of temporal odor concentration profile for plume tracking in dynamic turbulent environment. *IEEJ Transactions on Sensors and Micromachines* 138 (1): 15–22.

48 Muhtadi, M. and Nakamoto, T. (2018). Plume tracking strategy in turbulent environment using odor sensor with time constant. *Sensors and Materials* 30 (9): 2009–2021.

10

Microwave Chemical Sensors

Hamida Hallil[1,2] and Corinne Dejous[1]

[1] Univ. Bordeaux, CNRS, IMS, UMR 5218, Bordeaux INP, F-33405 Talence, France
[2] CINTRA, CNRS/NTU/THALES, UMI 3288, Research Techno Plaza Singapore 637553, Singapore

10.1 Interests of Electromagnetic Transducer Gas Sensors at Microwave Frequencies

The arrival of the Internet of Things (IoT) and the explosion of the communicating objects market require the deployment of embedded communicating sensors, preferably at low cost, providing exploitable real-time information and that can be operated in various and remote environments. This chapter presents a review on transducers in the radiofrequency (RF) domain, based on the interaction of target chemical species with chemical materials or functionalized composites. A part is devoted to the study of wireless readout techniques used with this type of sensor. Finally, techniques based on multivariate data analysis and machine learning that are needed to improve the selectivity and smartness of chemical sensors are discussed.

10.2 Operating Principle

10.2.1 Electromagnetic Transducers

Electromagnetic transduction is based on monitoring the variation of the electrical characteristics of an electromagnetic wave during its propagation in a medium or a sensitive material in the presence of a target species. Depending on the wave

Smart Sensors for Environmental and Medical Applications, First Edition. Edited by Hamida Hallil and Hadi Heidari.
© 2020 The Institute of Electrical and Electronics Engineers, Inc.
Published 2020 by John Wiley & Sons, Inc.

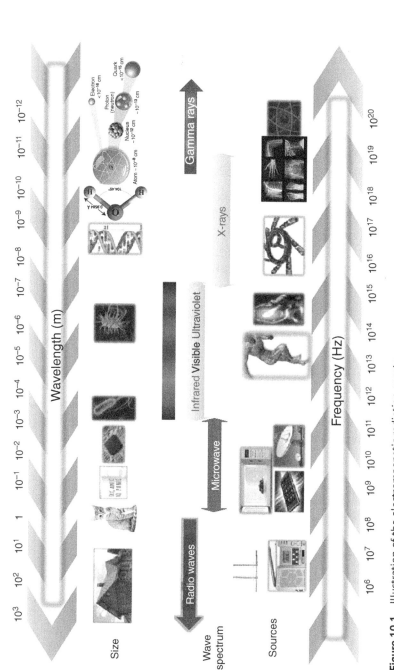

Figure 10.1 Illustration of the electromagnetic radiation spectrum.

frequency, the transduction is often optical or in the microwave range. The different radiations of the electromagnetic spectrum are presented in Figure 10.1.

By going from radio waves to gamma rays, the wavelength becomes shorter (the waves are more penetrating), whereas the frequency as well as the energy increases [1]. Table 10.1 presents some advantages and drawbacks of different electromagnetic transductions [2].

10.2.2 The Case of Microwave Transducers

The particular case of microwave transduction is based on the monitoring of the variation of electrical/dielectric properties (permittivity, ε; conductivity, σ; permeability,

Table 10.1 Some electromagnetic transducers advantages and drawbacks.

Transduction type	Advantages	Drawbacks	Frequency range
Optical	Ease of use in the absence of oxygen Insensitivity to interference with the electromagnetic environment	Sensitivity to ambient light interference Expensive, high-energy consumption Hardly integrable readout setup	From 4.3×10^{14} to 7.5×10^{14} Hz
Infrared	Physical technique (without sensitive layer) Use in inert atmospheres	Absence of infrared absorption by some gases Slow sequential monitoring Requires greater user expertise Expensive, high-energy consumption Hardly integrable readout setup	From 3×10^{11} to 4.3×10^{14} Hz
Microwave	Easy to design and implement at low cost Real-time detection Passive devices, possibly battery-less Possible remote readout Deployment in hostile environments	Sensitivity to electromagnetic interference and to environmental variations, circumvented with electromagnetic shielding or differential measurements	From 3×10^{8} to 3×10^{11} Hz

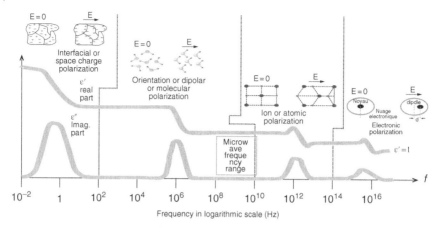

Figure 10.2 Evolution of the real and imaginary parts of the permittivity of water and polarization mechanisms.

μ) of a medium or sensitive material undergoing electromagnetic excitation with frequency within the microwave range, in a gaseous or liquid medium. The variations can be related to physical parameters such as temperature or pressure [3, 4] or to a sensitive layer (chemical sensors) [5, 6]. The dielectric permittivity is a complex quantity ($\varepsilon^* = \varepsilon' - \varepsilon''$) [7, 8], the most widely used electrical property in microwave transduction due to its link with the two other parameters (σ) and (μ) and to its strong variations with the frequency in the range of interest (Figure 10.2), associated to polarization phenomena [7].

Any interaction between the propagating electromagnetic wave and the target gas or physical quantity leads to a variation of the permittivity which results in an attenuation and/or a phase shift of the wave through the microwave transducer. Also, the microwave transducer is defined by its geometry which can be resistive, inductive and/or capacitive, or resonant. During its propagation, the electromagnetic wave (its field lines) interacts with its adjacent environment. Thus, it can be defined as effective permittivity (ε_{eff}) which varies according to the physical properties of the surrounding medium, such as temperature or pressure.

10.3 Theory of Microwave Transducers: Design, Methodology, and Approach

The design of this type of sensor is inspired by the geometries of electronic microwave components using transmission lines and not homogeneous guidance structures. Microlines offer possible miniaturization, 3D modeling, and low cost, unlike waveguides or volume structures [9]. Such planar lines have a two-dimensional structure; they allow the realization of passive or active RF circuits [7]. With at least two conductors, these structures allow a wave propagation without dispersion with a propagation constant equal to that of free waves in the substrate. This corresponds to the quasi-Transverse ElectroMagnetic (TEM) propagation mode, fundamental mode without cutoff frequency, as long as the operating frequency remains lower than the cut-off frequency of the first Transverse **Electric** (TE) or Transverse Magnetic (TM) mode. The quasi-TEM fields have a different phase velocity in the air and in the substrate, which prevents the continuity of their tangential component at the air–substrate interface in all points of the propagation axis and at every moment [10].

There is a variety of micro-transmission lines and each of them defines a technology. These include microstrip (MS) lines, striplines, slotlines, and coplanar waveguides (CPWs), each of which has its own specific geometrical and physical features [7].

As illustrated in Figure 10.3a, an MS transmission line consists of a conductive strip on one side of a dielectric or insulating substrate with a ground plane on the opposite face [11, 12]. Many studies have shown that a quasi-TEM mode propagates in such a line: the electric and magnetic fields are perpendicular to the axis of the transmission line, the field lines are mostly concentrated in the substrate between the metallized line and the ground plane (Figure 10.3b), a part of them being in the air surrounding the lines [13].

As shown in Figure 10.4, a CPW or transmission line consists of a hot line between two ground ribbons located on the same face of the substrate. With three conductive lines, two fundamental modes can propagate: the even mode is quasi-TE dispersive, and the odd mode is quasi-dispersive quasi-TEM, its electric and magnetic field lines are shown in Figure 10.4b. Although both modes are used, the odd one is often preferred due to low dispersion.

The characteristic impedance (Z_0) of a transmission line depends on its dimensions and the nature of the substrate. These parameters are linked by equations that are used to calculate each other [11–13]. Though such passive elements are

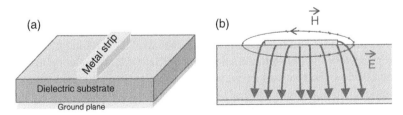

Figure 10.3 (a) Microstrip line and (b) distribution of the electric (*E*) and magnetic (*H*) field lines.

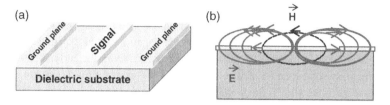

Figure 10.4 (a) Coplanar waveguide and (b) distribution of electric (*E*) and magnetic (*H*) field lines of the odd mode.

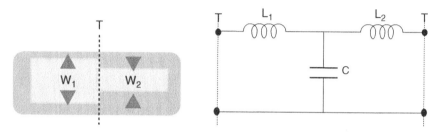

Figure 10.5 Discontinuity: junction at different width.

frequency-dependent, which is troublesome for applications such as filters, quasi-stationary passive elements (almost independent of the frequency) can be realized by using short lengths (i.e. lengths less than $\lambda_g/8$, smaller than $\lambda_g/10$ is recommended) [13]. A transmission line has a capacitance, an inductance, and a linear resistance per unit length. It can be modeled by a succession of π or T-filters like those presented in Figures 10.5 and 10.6. At a constant thickness, a wide path (low Z_0) has a capacitive behavior, while a thin line (high Z_0) has an inductive behavior.

Consisting of inductive, capacitive, and resistive localized or quasi-localized constants, MS components are commonly used in the microwave design, the choice depending mainly on the application (filter, transmission line, etc.), manufacturing technique, acceptable loss, quality factor (*Q*), power, and frequency of operation. These components are briefly described in Figures 10.7 and 10.8.

A quarter-wave line with a short-circuited end, behaves like a resonant circuit with elements in parallel, while a quarter-wave line open-ended, with inverted distributions of the current and the inverted voltage, behaves like a resonant circuit with elements in series. As a result, the short-circuited line behaves as an inductor at frequencies below the resonant frequency, as a capacitor at frequencies above the resonant frequency, and the converse behavior with the open-ended line.

To quantify this phenomenon, an electrical length of the line is defined such that the physical length (*l*) corresponds to a fraction of the guided wave period λ_g, equivalent to a phase θ of 360°, as illustrated in Figures 10.9 and 10.10.

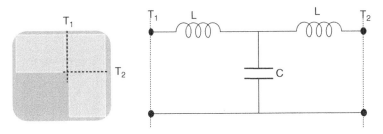

Figure 10.6 Discontinuity: bend at 90°.

Figure 10.7 Inductance with localized elements: (a) high impedance line, (b) meander line, (c) circular spiral line, (d) square spiral line, and (e) their ideal circuit representation.

Figure 10.8 Localized element capacity: (a) interdigitated, (b) metal–insulator–metal, and (c) their ideal circuit representation.

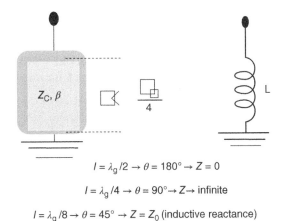

$$l = \lambda_g/2 \rightarrow \theta = 180° \rightarrow Z = 0$$

$$l = \lambda_g/4 \rightarrow \theta = 90° \rightarrow Z \rightarrow \text{infinite}$$

$$l = \lambda_g/8 \rightarrow \theta = 45° \rightarrow Z = Z_0 \text{ (inductive reactance)}$$

Figure 10.9 Stub short circuited.

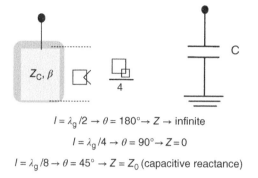

$$l = \lambda_g/2 \rightarrow \theta = 180° \rightarrow Z \rightarrow \text{infinite}$$
$$l = \lambda_g/4 \rightarrow \theta = 90° \rightarrow Z = 0$$
$$l = \lambda_g/8 \rightarrow \theta = 45° \rightarrow Z = Z_0 \text{ (capacitive reactance)}$$

Figure 10.10 Open stub.

10.4 Microwave Structure-Based Chemical Sensor

10.4.1 Manufacturing Techniques

Among advantages, such planar structures can be realized with additive manufacturing techniques. Furthermore, printed electronics is increasingly recognized as a key enabler for the IoT and part of the "Fourth Industrial Revolution" with functional and inexpensive technological advances [14, 15]. Indeed, printed electronics offer inexpensive and almost lossless manufacturing flexibility as it uses intelligent additive creation techniques often based on abundant and low-cost materials; suitably adapted processes that are environmentally friendly and mechanically flexible, makes them conformable and adaptable to conformable surfaces.

If multiple technologies or printing methods may be used in the field of printed electronics, they can be differentiated with respect to their ability to deposit on a substrate, a material as homogenous as possible to ensure a good conductivity, with the best possible resolution. Among these methods we can mention: screen printing, spin-coating, flexography, roll-to-roll, laser-induced forward transfer, and inkjet [15–20].

10.4.2 Chemical Microwave Sensors

To meet the growing need for simple, sensitive, portable, stable, communicating electronic sensors dedicated to the detection of various chemical molecules in a wide range of applications and with a view to reducing costs, research on nanomaterials for gas detection has enormously increased in recent years.

The desired nanomaterials for this type of communicating passive transducers, dedicated for applications such as IoTs, must have features such as a large specific surface area [21], long-term stability [22], high mobility of load carriers (allows low noise) [23], sensitivity at ambient temperature [24], selectivity that can be improved (functionalization) [25], compatibility with printable ink solutions, and integrability into low-temperature working protocols [26].

Among these nanomaterials we can mention carbon materials including graphene, carbon nanotubes (CNTs), and their composites such as PEDOT:PSS-MWCNT for poly (3,4-ethylenedioxythiophene) polystyrene sulfonate-multiwall carbon nanotubes which are used in several fields and applications, ranging from energy storage to biosensors [27, 28]. They are widely used for the detection of volatile organic compounds (VOCs) [25, 29, 30], hydrogen sulfide (H_2S) [25, 31], carbon dioxide (CO_2) [25, 26], ammonia (NH_3) [25, 32, 33], and nitrogen dioxide (NO_2) [24, 25, 34, 35].

Microwave sensors developed today combine such materials and obey the same rules as passive elements: filters, resonators, antennas, etc. [36]. Table 10.2 presents some examples of research work, on microwave transducers, which have been conducted in the past decade. It exhibits the following parameters: technology/geometry, operation frequency, sensitive material, target molecule and its concentration, sensitivity and measurement condition: static (S–Parameters as function of the frequency) or dynamic (Variation of S-Parameters and Frequency Shift as function of the time) [37–53]. These studies show the interest of this kind of transducers in terms of passivity, and use at ambient temperature, consequently with lower energy consumption. Working at high-frequency and possibly wirelessly interrogated, they are very appropriate for networking and communicating operation, being usable for real-time detection and providing exploitable information directly. In addition, based on a planar structure, the device can be manufactured on a flexible substrate by low-cost inkjet printing technology [54, 55].

Bailly et al. used a geometry based on conductor-backed coplanar waveguide (CBCPW) microwave antenna with a hematite sensitive layer, and demonstrated the high sensitivity for ammonia at concentrations ranging from 100 to 500 ppm [47]. This study shows that the microwave based on rhombohedra recorded the highest sensitivity with 4.6×10^{-7} ppm^{-1} and 3.4×10^{-7} ppm^{-1} for the real and imaginary parts of |ΔΓ|, respectively. Very recently, Park et al. also published a study based on a double split-ring resonator (DSRR) as microwave resonator coupled with PEDOT:PSS conducting polymer film for humidity detection [49]. They show that as the RH changes from 40 to 60%, the amplitude and the resonance frequency on the transmission parameter S21 change simultaneously and the sensor exhibits a great repeatability. At 40% RH, the magnitude and resonance frequency shifts in the transmission parameter S21 are estimated to 0.07 dB and −8.15 MHz. In the two last studies [51, 52], the structures deal with a differential dynamic study by integrating into the same platform two resonators. The first

Table 10.2 Some examples of microwave transducing sensors presented in the literature.

Technology/ geometry	Operating frequency (GHz)	Sensitive material	Target molecule	Concentrations	Sensitivity	Mode	References
Planar microwave active-resonator	1.52	Zeolite	CO_2	>45 vol.% CO_2	24 kHz/% CO_2	Static	[37]
Planar microstrip ring resonator	8.5	Zeolite	H_2O and ammonia	≤2 vol.% H_2O and 500 ppm NH_3	—	Dynamic	[38]
Flexible coupled microwave ring resonator	4	VOC polymeric adsorbent beads (V_5O_3)	Methyl ethyl ketone (MEK) and cyclohexane	250–1000 ppm	MEK: 40 kHz/ppm Cyclohexane: 2 kHz/ppm for	Static	[39]
Substrate integrated resonator (SIW)	4 and 6	Without	H_2	2% H_2	8.1 and 33.9%	Static	[40]
Coplanar structures	2 and 10	Zeolite	VOC: toluene	50–500 ppm	—	Dynamic	[41]
Microstrip inter digital capacitor	2.28 and 6.55	TiO_2	NH_3	100–500 ppm	—	Static/ dynamic	[42]
Substrate integrated cavity resonator	3.6 and 4.15	Without	H_2O	0–80% H_2O	101 kHz/RH	Static	[43]
Variable attenuator or phase shifter	2.4	Conducting polymer	VOC: ethanol	100 ppm	—	Dynamic	[44]
Flexible patch antenna	4.5	SWNT	Ammonia	50–100 ppm	43–49 MHz from 50–100 ppm	Static	[45]

Structure	Frequency (GHz)	Material	Gas	Range	Response	Static/dynamic	Ref.
Split ring resonator (SRR)	2.5	CNTs	Ammonia	Far and close gas exposure	20 MHz for far away gas 60 MHz shift for close gas exposure	Static	[46]
Conductor-backed coplanar waveguide (CBCPW) microwave antenna	5	Hematite α-Fe_2O_3 pseudocubes, rhombohedra, and spindle-like particles	Ammonia	100–500 ppm	Imaginary part of response sensor versus ppm	Static	[47]
LC structure	0.177	MoS_2 nanoflakes	RH %	10–60% RH	2.79 kHz/% RH	Static	[48]
Double split-ring resonator (DSRR)	2.55–2.95	PEDOT: PSS	RH %	40–60%	$\Delta S21$: 0.04–0.1 dB ΔF: −7 to −15 MHz	Static / dynamic	[49]
Microstrip line structure	1–10	PEDOT: PSS	RH %	10–50%	Resistance decreased by 6.43%	Static	[50]
A negative resistance oscillator	0.9	PEDOT: PSS	RH %	20–80%	ΔF: 4.2 MHz	Static/ dynamic	[51]
Flexible capacitive resonator bandpass filter	2.5 and 4.85	PEDOT: PSS MWCNTs	VOC: ethanol	500–2000 ppm	−2.482 KHz ppm^{-1}	Static/ dynamic	[52]
Flexible stub-based microwave resonator	0.6, 1.7, and 2.91	PEDOT: PSS MWCNTs	VOC: ethanol	500–2000 ppm	0.646 kHz ppm^{-1}	Static/ dynamic	[53]

resonator without sensitive material is the reference channel and therefore considered as a physical sensor, while the second resonator with sensitive material, based on PEDOT:PSS-MWCNTs conducting polymer, is referred as the sensitive channel (chemical sensor for target molecules). The differential measurement reduces the influence of physical effects such as pressure or temperature, the electromagnetic wave being very sensitive to its propagation environment. Different resonator structures and sensitive patterns, printed by ink jet on different substrates (paper, Kapton®), are studied. The tests carried out under a controlled gas atmosphere (ethanol and toluene in nitrogen) showed variations of amplitude and phase of the dynamic response as a function of the type and concentration of gas, at room temperature. These two studies present sensitivities of -2.482 kHz ppm^{-1} [52] and 0.646 kHz ppm^{-1} [53] to ethanol in the range from 0 to 1300 ppm, respectively, with sensors based on capacitive resonator-based bandpass filter (Figure 10.11a) and stub-based microwave resonator (Figure 10.11b). For example, Figure 10.12 illustrates the effect of the ethanol vapor on the resonant frequencies of the resonators according to the transmission parameters (S21) in static (Figure 10.12a) and real-time (Figure 10.12b) detection (dynamic).

10.4.3 Wireless Interrogation Schemes

The passive wireless microwave sensor operates based on the microwave backscatter phenomenon. The sensor integrates a detection unit with an antenna: the detection unit can be a microwave transmission line, a resonator, a filter, or an antenna structure. The antenna receives a broadband incident microwave signal from a reader, the sensor unit captures the frequency component with a value identical to its resonant frequency while reflecting the other frequency components of the incident signal. From the reflection characteristics, the resonant frequency

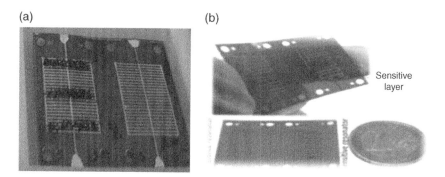

Figure 10.11 Manufactured devices: (a) capacitive resonator bandpass filter structure and (b) stub-based microwave resonator.

(a)

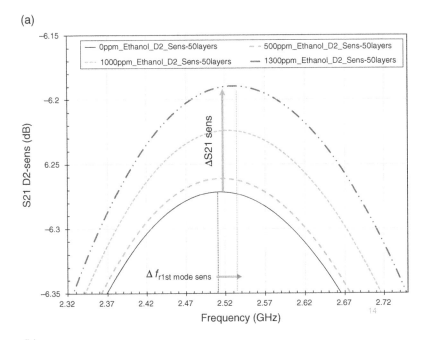

(b)

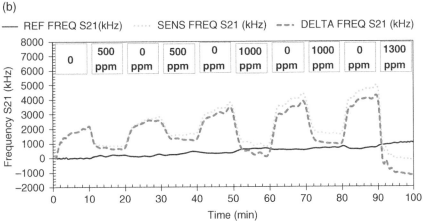

Figure 10.12 (a) Static evolution of S21 parameters of microwave sensor based on 50 layers of PEDOT:PSS-MWCNTS. (b) Dynamic evolution of the resonant frequency for different concentrations of ethanol (0–1300 ppm).

of the sensor can be extracted. As the surrounding medium and/or the sensitive layer on the sensor is modified, the electromagnetic wave propagating parameters and thus the backscattering characteristics are shifted in turn [56–58]. Many criteria will have a significant impact on the choice of the reader architecture and the sensitivity of the communication in a usually harsh environment [5, 59, 60]:

- Sensor backscatter detection technology: This is the only satisfactory technique for large distances (>10 m) and passive sensor technology with no power supply nor energy harvesting circuit
- Low echo of structures at distances greater than 20 m
- Microwave frequencies: The operating frequencies of microwave sensors are around a few GHz and are shifted by a few hundred kHz or even MHz in the presence of target species
- Application-specific signal processing: improve signal-to-noise level
- Sensor identification in a wireless network
- Easily usable in harsh environments

Currently, the techniques presented in the literature are based on readers of a passive RFID system, pulse interrogation or frequency modulation (FM), and radar technology reader [5, 60–62].

The reader of a passive RFID system has two functions. The first function is to realize the communication and the second to shape and analyze the signals coming from the sensor [25]. The signal processing part and the communication management are provided by a digital control unit generally composed of a microprocessor, a memory, and the control interface of the communication part [62, 63]. The communication is carried out by the analog part of the reader [64], with a transmitter and a receiver to manage a continuous wave (CW) carrier, sending it to the sensor and processing the echo signal. For authentication issue with RFID tags, the carrier is often mixed with an additional coded signal; the echo is then demodulated and processed. Though very attractive, these RFID systems suffer from a small interrogation distance, less than 10 m, and use quite low-frequency bands [65, 66].

When a resonant sensor is connected to an antenna, a pulse signal can excite it and generate an oscillation. The burst frequency is close to that of the interrogated resonator [67]. When the interrogation stops, a decreasing oscillation undergoes the modifications inside the sensor due to the environment, sent back to the reader via the antenna connected to its terminals. The signal received by the reader, mixed with the signal of a local oscillator is converted into frequency by an FFT. The spectral signal at the reader output is a series of lines offset by a frequency representative of the sensor shift, thus of the initial shift of the physical/chemical parameter. The pulse communication offers a low interrogation time (less than 10 μs) but the measurement resolution is often small.

The interrogation of the passive sensors is also carried out in the frequency domain by an emission of a continuous frequency-modulated signal, characterized by a frequency variation ΔF over a period T [68]. The signal reflected by the sensor antenna contains the delayed copies of the emitted signal, the delay being impacted by the sensor characteristics. The signal received by the reader is then converted into baseband by mixing with the original modulated signal. Unlike pulse interrogation, frequency-modulated interrogation offers a much better measurement resolution, as based on a delay, but the time of the communication is longer (T of some ms).

Radars are used in many applications such as level measurement, obstacle detection, meteorology, military, etc. They are mostly based on pulsed or CW technologies including FMCW FM [5, 60–70]. Pulse radars are limited in their ability to quickly and accurately measure the spectral components of a complex movement or the position of multiple targets concentrated in a restricted region such as in a sensor array [60]. In the family of CW radar, the FMCW radar transmits and receives a signal continuously, but the frequency of the transmitted signal is changed as a function of time and therefore, the frequency of the received signal can be used to measure of a delay. This FM allows the radar to determine the distance but also the speed of the target over a wide range. This separating ability is defined by its angular spatial resolution and its depth resolution.

In the study conducted by Hallil et al. [71], a FMCW radar was developed for the interrogation of a passive gas sensor using the structure (sensor) and antenna modes to differentiate the antenna and its sensor connected by a cable considered as a delay line. By this principle, the measurement of the sensor is optimal and note also that different line lengths allow to identify several sensors with adjacent spectral lines on the radar beat signal. This identification technique is useful for measuring a physical quantity regardless of the interrogation distance. For example, a sensor and its reference (pressure or gas detection) are connected to a single antenna by two lines of different lengths. The comparison of the radar detection levels of the two lines, gave the value of the two devices, imaging the two parameters, for the full scale of the interrogation range [5, 60].

These microwave sensor interrogation systems have paved the way for the development of high-frequency, long-range communication and for compact and fully passive measurement cells. However, to effectively and fully interrogate passive wireless sensors, there are some fundamental challenges to be faced, among them especially the environmental interferences that can override the backscattered detection signal.

10.5 Multivariate Data Analysis and Machine Learning for Targeted Species Identification

Multivariate data analysis and machine learning are emerging areas that offer performance and cost advantages for extracting valuable information from raw data sets. These techniques can perform exploratory and predictive analysis, which can help uncover hidden trends in the data resulting from the variations in the physicochemical properties of the transducers used as descriptive variables of the target species [72, 73].

The electronic nose and tongue are electronic systems imitating olfactory and taste biological functions to identify simple or complex odors or tastes [74–77]. Figure 10.13 shows a typical e-nose system. Indeed, when the volatile chemical molecules pass through the sensor array that is composed of different types of gas sensors, the sensor response will be recorded as "fingerprints" of the data. Beyond the initial acquisition of raw data by the sensor array, most of the data analysis and pattern recognition steps are typically accomplished through mathematical treatment of the data. Pattern analysis constitutes a critical building block in the development of gas sensor arrays that are designed for the detection, identification, and measurement of various gas species of interest. In order to design a robust pattern analysis system for e-nose, various issues involved in processing multivariate data need to be considered. These include signal processing and manipulation, multivariate analysis for feature extraction, classification and clustering of obtained features, and finally inference and decision making.

Such pattern analysis is involved in electronic nose technology, which offers a fast and nondestructive alternative for the detection of volatile complex odors or other chemical molecules, recognized as one of the best strategies in different fields such as the food industry, environmental monitoring, or bio-medical [78–80].

Among the most commonly used methods for multivariate data analysis, we can cite: Principal Component Analysis (PCA) [81] and Discriminant Analysis Algorithms, such as Linear Discriminant Analysis (LDA) [82], Partial Least-Squares discriminant analysis (PLS-DA) [83], and Orthogonal Partial Least-Squares Discriminant Analysis (OPLS-DA) [84]. PCA is especially a widely used method for extracting information from datasets, very popular and used with chemometrics that is widely deployed in the agri-food sector. This technique

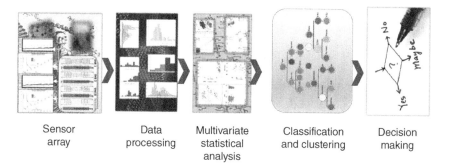

| Sensor array | Data processing | Multivariate statistical analysis | Classification and clustering | Decision making |

Figure 10.13 Typical e-nose system.

permits to visualize samples, often characterized by many descriptive measurements, in two-dimensional spaces formed of principal components, which are linear combinations of the original variables explaining the highest variance in the data. However, though PCA is known as a powerful visualization tool, it does not perform classification and cannot generalize information for new or unknown samples. A common method for that aim, is based on LDA to identify the most promising variables in terms of hidden models and use them as inputs for discriminant analysis models. The PCA–LDA approach remains widely used as a simple and rapid method of classifying samples [85].

In the case of machine learning-derived classification models, they combine concepts of multivariate data analysis, artificial intelligence, mathematical optimization, and computer science to efficiently learn and extract useful information from large and complex datasets and to predict information for new and unknown data samples. Some popular machine learning algorithms are Artificial Neural Networks (ANN), decision and regression trees, regression analysis, and Support-Vector Machines (SVM models, multilayer perceptron [MLP], and random decision forests) [86–88]. They have been used successfully in the recent literature to solve prediction problems in various fields and especially in the food industry. However, these algorithms are designed to work better with a large amount of data and are usually "black boxes," difficult to interpret as relationships between descriptors and their behaviors are not visible for most algorithms. These drawbacks may explain the relatively low popularity of machine learning techniques compared to traditional techniques of statistical and multivariate analysis allowing good discrimination of target species and from a reasonable number of data.

10.6 Conclusion and Prospects

The advantage of microwave frequency transduction is to process a signal compatible with wireless transmission without the need for frequency transposition, thereby reducing the consumption generated by this function and potentially leading to the integration of a completely passive sensor array. On the other hand, these components are also manufactured by additive technologies, compatible with the large-scale production of low-cost sensors, easy to integrate as they are conformable.

In addition, the proposed sensor ultimately provides information that can be directly exploited when it is integrated with its instrumentation system (wired or wireless interrogation), thus constituting a platform adapted for embedded applications and dedicated to the IoT. This opens very important prospects for the proliferation of detection and control sites with the realization and deployment of networks of wireless communicating sensors, even in dangerous and/or difficult to access environments.

With the current technological advances in data collection, storage, and processing, it is therefore natural that robust solutions such as combinations of multivariate data analysis and automatic learning methods are associated to chemical sensor arrays to improve their performance in terms of selectivity and identification of target species in complex environments. These methods are widely and successfully used for classification in several fields and exhibit high performance.

Acknowledgments

The authors are grateful for the financial supports of the French National Research Agency under Project ANR-13-BS03-0010 and the "Investments for the future" Programme IdEx Bordeaux, under Grant ANR-10-IDEX-03-02.

References

1 Wait, J.R. (1985). *Electromagnetic Wave Theory*, 195–196. New York: Harper & Row.
2 Awang, Z. (2014). Gas sensors: a review. *Sensors & Transducers* 168 (4): 61–75.
3 Aubert, H., Chebila, F., Jatlaoui, M. et al. (2013). Wireless sensing and identification based on radar cross section variability measurement of passive electromagnetic sensors. *Annals of Telecommunications – Annales des Télécommunications* 68 (7-8): 425–435.
4 Chahadih, A., Cresson, P.Y., Hamouda, Z. et al. (2015). Microwave/microfluidic sensor fabricated on a flexible kapton substrate for complex permittivity characterization of liquids. *Sensors and Actuators A: Physical* 229: 128–135.
5 Bernou, C., Rebière, D., and Pistre, J. (2000). Microwave sensors: a new sensing principle. Application to humidity detection. *Sensors and Actuators B: Chemical* 68 (1-3): 88–93.
6 Hallil, H., Bahoumina, P., Pieper, K. et al. (2019). Differential passive microwave planar resonator-based sensor for chemical particle detection in polluted environments. *IEEE Sensors Journal* 19 (4): 1346–1353.
7 Hallil, H. (2010). Conception et réalisation d'un nouveau capteur de gaz passif communicant à transduction RF. Doctoral dissertation, Université Paul Sabatier-Toulouse III.
8 Weir, W.B. (1974). Automatic measurement of complex dielectric constant and permeability at microwave frequencies. *Proceedings of the IEEE* 62 (1): 33–36.
9 Borazjani, O. and Rezaee, A. (2012). Design, simulation and construction a low pass microwave filters on the micro strip transmission line. *International Journal of Computer Theory and Engineering* 4 (5): 784.

10 Li, K., Kurita, D., & Matsui, T. (2005, June). An ultrawideband bandpass filter using broadside-coupled microstrip-coplanar waveguide structure. In *IEEE MTT-S International Microwave Symposium Digest*, 2005. (pp. 4-pp). IEEE.

11 Garg, R., Bahl, I., and Bozzi, M. (2013). *Microstrip Lines and Slotlines*. Artech House.

12 Schneider, M.V. (1969). Microstrip lines for microwave integrated circuits. *Bell System Technical Journal* 48 (5): 1421–1444.

13 Hong, J.-S.G. and Lancaster, M.J. (2004). *Microstrip Filters for RF/Microwave Applications*. Wiley.

14 Chang, J.S., Facchetti, A.F., and Reuss, R. (2017). A circuits and systems perspective of organic/printed electronics: review, challenges, and contemporary and emerging design approaches. *IEEE Journal on Emerging and Selected Topics in Circuits and Systems* 7 (1): 7–26.

15 Rosa, P., Câmara, A., and Gouveia, C. (2015). The potential of printed electronics and personal fabrication in driving the internet of things. *Open Journal of Internet of Things (OJIOT)* 1 (1): 16–36.

16 Haase, K., Hambsch, M., Da Rocha, C.T. et al. (2019). Advances in solution processing of organic materials for devices. In: *Handbook of Organic Materials for Electronic and Photonic Devices*, 551–577. Woodhead Publishing.

17 Abbel, R., Galagan, Y., and Groen, P. (2018). Roll-to-roll fabrication of solution processed electronics. *Advanced Engineering Materials* 20 (8): 1701190.

18 Ngo, T.D., Kashani, A., Imbalzano, G. et al. (2018). Additive manufacturing (3D printing): a review of materials, methods, applications and challenges. *Composites Part B: Engineering* 143: 172–196.

19 Qian, R.-C. and Long, Y.-T. (2018). Wearable chemosensors: a review of recent progress. *ChemistryOpen* 7 (2): 118–130.

20 Smits, E.C.P., Walter, A., De Leeuw, D.M. et al. (2017). Laser induced forward transfer of graphene. *Applied Physics Letters* 111 (17): 173101.

21 Varghese, S.S., Lonkar, S., Singh, K.K. et al. (2015). Recent advances in graphene based gas sensors. *Sensors and Actuators B: Chemical* 218: 160–183.

22 Eshkalak, S.K., Chinnappan, A., Jayathilaka, W.A.D.M. et al. (2017). A review on inkjet printing of CNT composites for smart applications. *Applied Materials Today* 9: 372–386.

23 Lakkis, S., Younes, R., Alayli, Y. et al. (2014). Review of recent trends in gas sensing technologies and their miniaturization potential. *Sensor Review* 34 (1): 24–35.

24 Huang, L., Wang, Z., Zhang, J. et al. (2014). Fully printed, rapid-response sensors based on chemically modified graphene for detecting NO_2 at room temperature. *ACS Applied Materials & Interfaces* 6 (10): 7426–7433.

25 Wang, T., Huang, D., Yang, Z. et al. (2016). A review on graphene-based gas/vapor sensors with unique properties and potential applications. *Nano-Micro Letters* 8 (2): 95–119.

26 Rana, M.M., Ibrahim, D.S., Mohd Asyraf, M.R. et al. (2017). A review on recent advances of CNTs as gas sensors. *Sensor Review* 37 (2): 127–136.

27 Sun, K., Zhang, S., Li, P. et al. (2015). Review on application of PEDOTs and PEDOT: PSS in energy conversion and storage devices. *Journal of Materials Science: Materials in Electronics* 26 (7): 4438–4462.

28 Hui, Y., Bian, C., Xia, S. et al. (2018). Synthesis and electrochemical sensing application of poly(3,4-ethylenedioxythiophene)-based materials: a review. *Analytica Chimica Acta* 1022: 1–19.

29 Jeon, J.-Y., Kang, B.-C., Byun, Y.T. et al. (2019). High-performance gas sensors based on single-wall carbon nanotube random networks for the detection of nitric oxide down to the ppb-level. *Nanoscale* 11 (4): 1587–1594.

30 Ahmad, A., Lokhat, D., Setapar, S.H.M. et al. (2019). Nanocarbon composites for detection of volatile organic compounds. In: *Nanocarbon and Its Composites*, 401–419. Woodhead Publishing.

31 Asad, M., Sheikhi, M.H., Pourfath, M. et al. (2015). High sensitive and selective flexible H_2S gas sensors based on Cu nanoparticle decorated SWCNTs. *Sensors and Actuators B: Chemical* 210: 1–8.

32 Abdelhalim, A., Falco, A., Loghin, F. et al. (2016). Flexible NH_3 sensor based on spray deposition and inkjet printing. In: *2016 IEEE SENSORS*, 1–3. IEEE.

33 Eising, M., Cava, C.E., Salvatierra, R.V. et al. (2017). Doping effect on self-assembled films of polyaniline and carbon nanotube applied as ammonia gas sensor. *Sensors and Actuators B: Chemical* 245: 25–33.

34 Lee, S.W., Lee, W., Hong, Y. et al. (2018). Recent advances in carbon material-based NO_2 gas sensors. *Sensors and Actuators B: Chemical* 255: 1788–1804.

35 Choi, S.-W., Kim, J., and Byun, Y.T. (2017). Highly sensitive and selective NO_2 detection by Pt nanoparticles-decorated single-walled carbon nanotubes and the underlying sensing mechanism. *Sensors and Actuators B: Chemical* 238: 1032–1042.

36 Hallil, H., Menini, P., and Aubert, H. (2009). Novel microwave gas sensor using dielectric resonator with SnO_2 sensitive layer. *Procedia Chemistry* 1 (1): 935–938.

37 Zarifi, M.H., Gholidoust, A., Abdolrazzaghi, M. et al. (2018). Sensitivity enhancement in planar microwave active-resonator using metal organic framework for CO_2 detection. *Sensors and Actuators B: Chemical* 255: 1561–1568.

38 Bogner, A., Steiner, C., Walter, S. et al. (2017). Planar microstrip ring resonators for microwave-based gas sensing: design aspects and initial transducers for humidity and ammonia sensing. *Sensors* 17 (10): 2422.

39 Abbasi, Z., Zarifi, M.H., Shariati, P. et al. (2017). Flexible coupled microwave ring resonators for contactless microbead assisted volatile organic compound detection. In: *2017 IEEE MTT-S International Microwave Symposium (IMS)*, 1228–1231. IEEE.

40 Ndoye, M., EL Matbouly, H., Sama, Y.N. et al. (2016). Sensitivity evaluation of dielectric perturbed substrate integrated resonators for hydrogen detection. *Sensors and Actuators A: Physical* 251: 198–206.

41 De Fonseca, B., Rossignol, J., Bezverkhyy, I. et al. (2015). Detection of VOCs by microwave transduction using dealuminated faujasite DAY zeolites as gas sensitive materials. *Sensors and Actuators B: Chemical* 213: 558–565.

42 Bailly, G., Harrabi, A., Rossignol, J. et al. (2016). Microwave gas sensing with a microstrip interDigital capacitor: detection of NH_3 with TiO_2 nanoparticles. *Sensors and Actuators B: Chemical* 236: 554–564.

43 El Matbouly, H., Boubekeur, N., and Domingue, F. (2015). Passive microwave substrate integrated cavity resonator for humidity sensing. *IEEE Transactions on Microwave Theory and Techniques* 63 (12): 4150–4156.

44 Lee, Y.-J., Kim, B.-H., Lee, H.-J. et al. (2014). A reflection type gas sensor using conducting polymer as a variable impedance at microwave frequencies. In: *2014 IEEE SENSORS*, 1819–1822. IEEE.

45 Lee, H., Shaker, G., Naishadham, K. et al. (2011). Carbon-nanotube loaded antenna-based ammonia gas sensor. *IEEE Transactions on Microwave Theory and Techniques* 59 (10): 2665–2673.

46 Pardue, C., Naishadham, K., Song, X. et al. (2014). Integration of carbon nanotube films with SRRs for air quality sensing applications. In: *WAMICON 2014*, 1–3. IEEE.

47 Bailly, G., Rossignol, J., De Fonseca, B. et al. (2016). Microwave gas sensing with hematite: shape effect on ammonia detection using pseudocubic, rhombohedral, and spindlelike particles. *ACS Sensors* 1 (6): 656–662.

48 Su, S., Lv, W., Zhang, T. et al. (2018). A MoS_2 nanoflakes-based LC wireless passive humidity sensor. *Sensors* 18 (12): 4466.

49 Park, J.-K., Kang, T.-G., Kim, B.-H. et al. (2018). Real-time humidity sensor based on microwave resonator coupled with PEDOT: PSS conducting polymer film. *Scientific Reports* 8 (1): 439.

50 Kang, T. G., Park, J. K., Kim, B. H., Lee, J. J., Choi, H. H., Lee, H. J., & Yook, J. G. (2019). Microwave characterization of conducting polymer PEDOT: PSS film using a microstrip line for humidity sensor application. *Measurement*, 137, 272–277.

51 Kang, T.-G., Park, J.-K., Yun, G.-H. et al. (2019). A real-time humidity sensor based on a microwave oscillator with conducting polymer PEDOT: PSS film. *Sensors and Actuators B: Chemical* 282: 145–151.

52 Bahoumina, P., Hallil, H., Lachaud, J.-L. et al. (2018). Chemical sensor based on a novel capacitive microwave flexible transducer with polymer nanocomposite-carbon nanotube sensitive film. *Microsystem Technologies*: 1–14. https://doi.org/10.1007/s00542-018-4099-4AR8.

53 Bahoumina, P., Hallil, H., Lachaud, J.-L. et al. (2017). Microwave flexible gas sensor based on polymer multi wall carbon nanotubes sensitive layer. *Sensors and Actuators B: Chemical* 249: 708–714.

54 Vyas, R., Lakafosis, V., Lee, H. et al. (2011). Inkjet printed, self powered, wireless sensors for environmental, gas, and authentication-based sensing. *IEEE Sensors Journal* 11 (12): 3139–3152.

55 Vena, A., Sydänheimo, L., Tentzeris, M.M. et al. (2015). A fully inkjet-printed wireless and chipless sensor for CO_2 and temperature detection. *IEEE Sensors Journal* 15 (1): 89–99.

56 Alemdar, H. and Ersoy, C. (2010). Wireless sensor networks for healthcare: A survey. *Computer Networks* 54 (15): 2688–2710.

57 Alreshaid, A.T., Hester, J.G., Su, W. et al. (2018). Ink-jet printed wireless liquid and gas sensors for IoT, SmartAg and Smart City applications. *Journal of the Electrochemical Society* 165 (10): B407–B413.

58 Zhang, J., Sunny, A.I., Zhang, G. et al. (2018). Feature extraction for robust crack monitoring using passive wireless RFID antenna sensors. *IEEE Sensors Journal* 18 (15): 6273–6280.

59 Chen, X., Wang, X., Li, H., and Xiong, J. (2018). Microwave based wireless passive sensor: a promising candidate for harsh environment application. In: *2018 IEEE Asia-Pacific Conference on Antennas and Propagation (APCAP)*, 262–263. IEEE.

60 Chebila, F. (2011). Lecteur radar pour capteurs passifs à transduction radio fréquence. Doctoral dissertation. Institut National Polytechnique de Toulouse (INP Toulouse)

61 Bariya, M., Nyein, H.Y.Y., and Javey, A. (2018). Wearable sweat sensors. *Nature Electronics* 1 (3): 160.

62 Karmakar, N.C., Amin, E.M., and Saha, J.K. (2016). *Chipless RFID Sensors*. Wiley.

63 Preradovic, S. and Karmakar, N.C. (2006). RFID readers-a review. In: *2006 International Conference on Electrical and Computer Engineering*, 100–103. IEEE.

64 Preradovic, S. and Karmakar, N.C. (2009). Design of short range chipless RFID reader prototype. In: *2009 International Conference on Intelligent Sensors, Sensor Networks and Information Processing (ISSNIP)*, 307–312. IEEE.

65 Ukkonen, L., Sydänheimo, L., and Kivikoski, M. (2007). Read range performance comparison of compact reader antennas for a handheld UHF RFID reader. In: *2007 IEEE International Conference on RFID*, 63–70. IEEE.

66 Hartmann, C.S. and Claiborne, L.T. Fundamental limitations on reading range of passive IC-based RFID and SAW-based RFID. In: *2007 IEEE International Conference on RFID*, vol. 2007, 41–48. IEEE.

67 Kalinin, V. (2001). Modelling of a wireless SAW system for multiple parameter measurement. In: *Proceedings of the IEEE Ultrasonics Symposium*, 1790–1793. IEEE.

68 Wolff, U., Schmidt, F., Scholl, G. et al. (1996). Radio accessible SAW sensors for non-contact measurement of torque and temperature. In: *1996 IEEE Ultrasonics Symposium. Proceedings*, 359–362. IEEE.

69 Tiuri, M. (1987). Microwave sensor applications in industry. In: *1987 17th European Microwave Conference*, 25–32. IEEE.

70 Stove, A.G. (1992). Linear FMCW radar techniques. In: *IEE Proceedings F (Radar and Signal Processing)*, 343–350. IET Digital Library.

71 Hallil, H., Chebila, F., Menini, P. et al. (2010). Feasibility of passive gas sensor based on whispering gallery modes and its RADAR interrogation: theoretical and experimental investigations. *Sensors & Transducers* 116 (5): 38.

72 Maione, C., Barbosa, F. Jr., and Barbosa, R.M. (2019). Predicting the botanical and geographical origin of honey with multivariate data analysis and machine learning techniques: a review. *Computers and Electronics in Agriculture* 157: 436–446.

73 Faust, O., Hagiwara, Y., Hong, T.J. et al. (2018). Deep learning for healthcare applications based on physiological signals: a review. *Computer Methods and Programs in Biomedicine* 161: 1–13.

74 Esteves, C.H.A., Iglesias, B.A., Ogawa, T. et al. (2018). Identification of tobacco types and cigarette brands using an electronic nose based on conductive polymer/porphyrin composite sensors. *ACS Omega* 3 (6): 6476–6482.

75 Al-Maskari, S., Xu, Z., Guo, W. et al. (2018). Bio-inspired learning approach for electronic nose. *Computing* 100 (4): 387–402.

76 Buratti, S., Malegori, C., Benedetti, S. et al. (2018). E-nose, e-tongue and e-eye for edible olive oil characterization and shelf life assessment: a powerful data fusion approach. *Talanta* 182: 131–141.

77 Pearce, T.C., Schiffman, S.S., Nagle, H.T., and Gardner, J.W. (eds.) (2006). *Handbook of Machine Olfaction: Electronic Nose Technology*. Wiley.

78 Du, D., Wang, J., Wang, B. et al. (2019). Ripeness prediction of postharvest kiwifruit using a MOS e-nose combined with chemometrics. *Sensors* 19 (2): 419.

79 Cui, S., Ling, P., Zhu, H. et al. (2018). Plant pest detection using an artificial nose system: a review. *Sensors* 18 (2): 378.

80 Saviauk, T., Kiiski, J.P., Nieminen, M.K. et al. (2018). Electronic nose in the detection of wound infection bacteria from bacterial cultures: a proof-of-principle study. *European Surgical Research* 59 (1–2): 1–11.

81 Chang, J.-E., Lee, D.-S., Ban, S.-W. et al. (2018). Analysis of volatile organic compounds in exhaled breath for lung cancer diagnosis using a sensor system. *Sensors and Actuators B: Chemical* 255: 800–807.

82 Xi, H., Li, X., Liu, Q. et al. (2018). Cationic polymer-based plasmonic sensor array that discriminates proteins. *Analyst* 143 (22): 5578–5582.

83 Lee, L.C., Liong, C.-Y., and Jemain, A.A. (2018). Partial least squares-discriminant analysis (PLS-DA) for classification of high-dimensional (HD) data: a review of contemporary practice strategies and knowledge gaps. *Analyst* 143 (15): 3526–3539.

84 Santos, F.A.d., Sousa, I.P., Furtado, N.A.J.C. et al. (2018). Combined OPLS-DA and decision tree as a strategy to identify antimicrobial biomarkers of volatile oils analyzed by gas chromatography–mass spectrometry. *Revista Brasileira de Farmacognosia* 28 (6): 647–653.

85 Sun, X., Liu, P., and Mancin, F. (2018). Sensor arrays made by self-organized nanoreceptors for detection and discrimination of carboxylate drugs. *Analyst* 143 (23): 5754–5763.

86 Cleophas, T.J. and Zwinderman, A.H. (2018). Regression trees. In: *Regression Analysis in Medical Research*, 359–364. Cham: Springer.

87 Ekins, S., Clark, A.M., Perryman, A.L. et al. (2018). Accessible machine learning approaches for toxicology. *Computational Toxicology: Risk Assessment for Chemicals*: 1–29.

88 Deist, T.M., Dankers, F.J.W.M., Valdes, G. et al. (2018). Machine learning algorithms for outcome prediction in (chemo) radiotherapy: an empirical comparison of classifiers. *Medical Physics* 45 (7): 3449–3459.

Index

Smart Sensors for Environmental and Medical Applications, First Edition. Edited by
Hamida Hallil and Hadi Heidari.
© 2020 The Institute of Electrical and Electronics Engineers, Inc.
Published 2020 by John Wiley & Sons, Inc.

IEEE Press Series on Sensors

Series Editor: Vladimir Lumelsky, Professor Emeritus, Mechanical Engineering,
University of Wisconsin-Madison

Sensing phenomena and sensing technology is perhaps the most common thread that connects just about all areas of technology, as well as technology with medical and biological sciences. Until the year 2000, IEEE had no journal or transactions or a society or council devoted to the topic of sensors. It is thus no surprise that the IEEE Sensors Journal launched by the newly-minted IEEE Sensors Council in 2000 (with this Series Editor as founding Editor-in-Chief) turned out to be so successful, both in quantity (from 460 to 10,000 pages a year in the span 2001–2016) and quality (today one of the very top in the field). The very existence of the Journal, its owner, IEEE Sensors Council, and its flagship IEEE SENSORS Conference, have stimulated research efforts in the sensing field around the world. The same philosophy that made this happen is brought to bear with the book series.

Magnetic Sensors for Biomedical Applications
Hadi Heidari, Vahid Nabaei
Smart Sensors for Environmental and Medical Applications
Hamida Hallil, Hadi Heidari

Printed and bound by CPI Group (UK) Ltd, Croydon, CR0 4YY